Fabricated Man

FABRICATED MAN

THE ETHICS OF GENETIC CONTROL

by Paul Ramsey

New Haven and London, Yale University Press, 1970

1. Genetic engineering--Moral and ethical aspects.
2. Medical ethics.

Library of Congress catalog card number: 78-123395
ISBN: 0-300-01373-6 (cloth), 0-300-01374-4 (paper)
Designed by Sally Sullivan,
set in IBM Selectric Press Roman type,
and printed in the United States of America by
The Carl Purington Rollins Printing-Office
of the Yale University Press.
Distributed in Great Britain, Europe, and Africa by
Yale University Press, Ltd., London; in Canada by
McGill-Queen's University Press, Montreal; in Mexico
by Centro Interamericano de Libros Académicos,
Mexico City; in Australasia by Australia and New
Zealand Book Co., Pty., Ltd., Artarmon, New South
Wales; in India by UBS Publishers' Distributors Pvt.,
Ltd., Delhi; in Japan by John Weatherhill, Inc.,
Tokyo.

TO
MY GRANDSON
JASON KAI COOPER

Contents

Chapter 1: Moral and Religious Implications of Genetic Control

The eugenic movement of the late nineteenth and early twentieth century was based in the main upon biological and socioscientific misinformation or lack of information, and—what is worse—upon parochial if not indeed elitist and racial views of the ideal type of man. An excellent history of this movement is Mark H. Haller's *Eugenics: Hereditarian Attitudes in American Thought.*[1] To read this book is to conclude, with R. S. Morison, Director, Division of Biological Sciences, Cornell University, that "the thing that has saved man from his limited visions in the past has been the difficulty of devising suitable means for reaching them."[2] The culmination or abuse of eugenics in the ghastly Nazi experiments would seem to be sufficient to silence forever proposals for genetic control.

However, this is not the case, and for two reasons. First, contemporary geneticists are increasingly being driven to varying degrees of gloom regarding the future of mankind because of the inexorable degeneration of the human genetic pool under the conditions of modern life. Secondly, since World War II molecular biology has steadily increased the range and precision of our knowledge of genetics. This may make it possible to determine some of the objectives of a program of genetic improvement or of at least a program for preventing further genetic deterioration, and also make it possible to devise suitable means for reaching these ends. Because of the ineluctable increase of the problem and of the knowledge that might afford some solution to it, it can safely be predicted that the future will see more rather than less discussion of proposals for genetic control.

Scientists will continue to debate these issues among themselves, and in the public forum. One scientist recently expressed a fitting sense of humility before the as yet unfathomed mystery of nature and nature's God, by asking his fellow scientists: "If any one of us had devised a mechanism as complex as the situation of the human race, how would we feel about letting any of our colleagues monkey about with it, on the assumption that they knew as little about it as we know about the psychosocial mechanism?"[3] Still this would be a weak foundation, and probably a vanishing foundation, on which to base opposition to genetic control. It is for the scientist to pay attention to objections of this sort, and to call them to the attention of the public. The present state of scientific knowledge, however, and the enormous practical obstacles in the way, ought not to be given important place in a discussion of the moral and religious issues raised by present and future eugenic proposals.

The Problem and the Issues

It is incumbent upon me, however, to describe in summary fashion mankind's problematic genetic situation as this is understood by certain contemporary geneticists. This would not be so bad if from generation to generation a more or less stable pool of genes were in passage, with its particular balance of physical, mental and emotional strengths or defects. The fact is, however, that in addition to the load of genetic deficiency from the previous generation, one out of every five persons now living (twenty percent) bears a deleterious mutation that has arisen with him and which he will pass on to or through any offspring he may have. The quality of human beings to be born could be maintained at its present level if, and only if, twenty percent of us become genetically extinct, either by failing to reach reproductive age or by not having children. The fact is that, because of our technical and medical competence and our proper concern for persons now alive, we are enabling people to reach the age of reproduction, and to reproduce when they do, in greater numbers than would have been the case in former ages.

Normal balancing elimination is not hereby frustrated; it is only postponed. One can be sure that some future generation will begin to experience twenty percent genetic deaths. But then it will be too late, since by then those who manage to stay alive will be generally and seriously impaired because of the genetic load they bear. Unless some conscious slection is made and some positive or negative direction given to generation, natural selection will, as it were, return to do this by more inhumane means than men could devise. Just as it will be too late if we do not adopt deliberate control of the *numbers* to be born and if we simply wait for overcrowding of the planet and starvation to correct overpopulation, so with respect to the *quality* or the mental and physical strength of the population generally, it will be too late (and indeed it will be inhumane) if we do not adopt measures to counteract the genetic deterioration which modern civilization and humanitarianism foster, and if we simply wait for balancing selection to overtake us and pull out the plug our hospitals now place in the way of the extinction of genetic defects.

Diabetes affords us a case in point. Diabetics formerly died early. After a cure was found in insulin, they were enabled to survive and lead useful lives. Since, however, they were not generally able to have children, these individuals were as genetically dead as if they had been stillborn. Now the safe delivery of the children of diabetic mothers is a commonplace in all our hospitals; and as a consequence the incidence of diabetes in the population is irreversibly increasing. "Of course we can get along with a lot more diabetics, and with good medical care they can live happily and bear diabetic children of their own. But there is a limit beyond which this process cannot be carried, and if we consider not diabetes alone, but all other ills to which the human race is genetically heir, that limit is not far away."[4]

The laws of genetics are general ones. They apply to all human endowments and disendowments. "This prospect is not pleasant to contemplate, but insulin injections may, conceivably, have to be as common in some remote future as taking aspirin tablets is at present."[5] We may be able to say of any one of the hereditary defects that can be singled out that, since (for example) eye glasses can remedy myopia, "the effort that would be needed to eradicate or reduce the frequency of myopia would exceed that requisite to rectify the defect environmentally."[6] We may even want to yield to the movement that has made wearing eye glasses romantically attractive, or at least not unattractive, thus feeding back rather than eliminating the genes for this weakness. But it is impossible to say this concerning the apparently irreversible increase of all the ills to which man is heir. This is the harsh reality the science of genetics discloses to us beneath the illusion (fostered by other sciences) that we are gradually conquering disease. What is good for the individual, and the generation now alive, is bad for the human race.

The geneticists I have read do not disagree as to the trend, but only concerning the degree or rapidity of the trend. In other words, they are pessimistic or optimistic within an ultimate genetic pessimism. Some say "the facts of human reproduction are all gloomy"[7] Others correct this by pointing out that medical science has mainly eliminated death before reproductive age from *infectious* diseases. These hold the opinion that a large fraction of *prenatal* and pre-adult deaths occur from recognizably genetic causes, and that genetic elimination is taking place at a good rate despite the rise in the longevity index.[8] There is also disagreement over whether mutations are always deleterious, and whether the genetic cause of serious defects may not also be the cause of "hybrid vigor" in the gene pool generally. But there is no disagreement about the trend, nor about the fact that there is an exponential increase of micro-genetic identification of genetic illnesses that could be either eliminated or greatly reduced in a generation or two.

Now, it may be objected that intellectual traits and emotional or moral capacities (in contrast to physical traits or capacities such as the color or the myopia of our eyes or our upright stature) are the product not of our genes, but of our environment or our choices and willpower, and that in any case they may be markedly improved or corrected only by the training supplied by our human or cultural environment. To this it has to be replied that hereditary physical diseases and defects are curable no less than infectious diseases, and that the plasticity of intellectual capacities and emotional proclivities to environmental changes constitutes no special case. There is here at most a difference of degree. The premise or the finding of the science of genetics is that there is a genetic basis of mental and moral traits no less than of physical traits, and that the laws of gene frequency and the processes of mutation and selection apply no less to the higher human attributes than they do to the lower.[9] It is important to remember that genetic determination ("determiners" might be a better word) means genetic endowment or disendowment. It does not establish a regime of inexorable necessity to which man is subject. The relationship between genotype and the environment is a dynamic one. The genotype determines not rigid traits of the organism (the phenotype) but rather its norm of reaction to the environment. The chief disagreement concerns whether we yet know enough, or can know enough, about the genetic basis of complex moral and mental characteristics to do anything about the transmission of defects, except in extreme cases.

Moreover, the conditions of modern life are producing a negative feedback in place of the positive feedback in strengths of mind and character which was the case during most of man's evolution. This is comparable to the negative feedback of physical defects modern medicine has produced.

Darwinian "fitness" means reproductive fitness.[10] Of course, evolution has bred into mankind a capacity to give blow for blow, because the individuals most adept at this thereby preserved their genes. But so do individuals who are willing to give help for help. We must, therefore, correct the gladiatorial image we have of the survival of the fittest. "The fittest, in the evolutionary sense, is nothing more spectacular than the quiet, often unobtrusive fellow who, rather than spend his time in combat, produces, feeds, and teaches a large family of children."[11]

Today, however, "everyone is helped to live according to his need and to reproduce according to his greed, or lack of foresight, skill, or scruple."[1][2] Cultural evolution no less than medical science has brought us around a turning point. For long ages before this point, man was becoming physically, mentally and morally fitter, but after it, man may be becoming not only physically, but mentally and morally *less* fit. "I think there is no question whatsoever," said Professor Ernst Mayr of Harvard, "that when there were smaller human groups, the selective premium on altruistic traits and cooperative traits that helped the survival of a well-integrated group must have been exceedingly high. Today, however, in a big metropolitan civilization, even highly antisocial behavior is not especially severely punished by natural selection."[13] As is his wont, H. J. Muller draws the more negative conclusion: "In fact, it seems not unlikely that in regard to the human faculties of the highest group importance—such as those needed for integrated understanding, foresight, scrupulousness, humility, regard for others, and self-sacrifice—cultural conditions today may be conducive to an actual lower rate of reproduction on the part of their possessors than of those with the opposite attributes."[14]

Because of the negative feedback of genes—producing poorer mental, emotional, and moral traits and suspending the selection of better traits—the crisis of our present-day civilization is genetic at least in part, and one that goes to the very *humanum* of man. This is an unfavorable, indeed a perilous, aspect of the fact that ours is now "one world," a fact which is ordinarily unqualifiedly praised as a consummation devoutly to be wished. As our common culture gradually becomes that of "one world," it extends to include the entire human gene pool. If there is operative a negative feedback of genes for physical, mental and moral traits alike, then the failure of this culture is to be expected (like the failures of all civilizations of the past). But this time the failure of the culture will mean failure for the entire human race (where before there were not only more or less encapsulated primitive tribes, but great gene pools to begin the next advance).

Thus, by doing away with the natural selection that used to keep us reasonably fit, by holding at bay the lethality of lethal genes, and by weakening the disfavor formerly placed upon bearers of unsociable traits, mankind is allowing an insidious genetic deterioration that will leave us less fit than when we began.

This brings us, finally, to the effect of radiation on mutation rates. The discussion in recent years of the hideous birth defects and monstrosities that would result for generations to come from a nuclear war (or from high exposure to atmospheric fallout from testing) can be put in proper perspective only when we remember that none of these genetic defects are new occurrences. They are only the result of accelerating the *rate* of mutation that is going on all the time in human sex cells and accumulating in the pool of genes. The genetic effect of radioactive fallout is small by comparison with the genetic dilemma of normal negative feedback.

It is also small by comparison with the consequences of the use of radiation energy for nonmilitary purposes that will be normal for centuries to come. Exposure to radioactive fallout is decidedly smaller than exposure to the medical and diagnostic use of radiation and the industrial use of atomic energy. The amount of radiation that will reach the reproductive cells from atmospheric testing is only about a hundreth as much per individual in the United States as that from medical diagnosis.[15] Let us assume that ten generations pass before the end of this age of nuclear energy, i.e. until mankind learns to make direct use of solar energy as before he took his power from the sun in the form of fossil fuels. The resulting exposure would lead to thirty million genetic deaths, if one foolishly calculated the number literally for time without end. On a more significant calculation, for, say, twenty generations or six hundred years, it would lead to 180,000 genetic deaths per generation or 6,000 annually—at the end of which span of time ninety percent of the induced mutants would still remain to be eliminated.[16] Plus, of course, a generally weakened population. Thus, on all counts the present age seems to be genetically debasing.

Now, "we are all fellow mutants together,"[17] and the question before us is: What are we going to do about our genetic dilemma? There is one and only one way of avoiding the "fiasco of a full fledged resumption of ordinary natural selection. That method, whether we like it or not, is purposive control over reproduction"[18]—over its quality no less than its quantity.

If there is a solution to the genetic problem, it will consist of finding some acceptable means of reducing the genetic load, or at least halting the increase of this load. To reduce the genetic load or to prevent further genetic deterioration, one would have to lower the mutation rate (or perform some operation that would cause back-mutation) and/or increase the elimination rate. This brings us to the two sorts of proposals for genetic control that are possible today or are envisioned as future possibilities.

(1) The first is some direct attack upon the deleterious mutated gene, either by what is called "genetic sur-

gery," "micro-surgery," or "nano-surgery,"[19] or by the introduction of some anti-mutagent chemical that will cause the gene to mutate back or will eliminate it from among the causes of genetic effects. At some time in the near or distant future this may be the means employed in a program of "negative" or "preventive" eugenics. Since a "bad" gene has to be replaced by a "good" one, this method could also be employed to induce or *direct* a program of "progressive" eugenics or "positive" genetic improvement.

(2) The second sort of available means arises from focusing attention upon the "phenotype" and not the "genotype": eugenically directed birth control, "parental selection," "germinal choice," or gross "empirical" selection for traits in human reproduction. The means available for accomplishing this could be used either in a program of "breeding in" desired traits ("progressive" or "positive" eugenics), or in a program of "breeding out" undesired traits ("preventive" or "negative" eugenics). Such means would be the control of conception by persons advised to do so by heredity clinics; voluntary sterilization; artificial insemination for eugenic reasons with semen from a nonhusband donor (AID), sometimes called "pre-adoption"; "foster pregnancy" or artificial inovulation, which is the reverse of AID and may be arranged so as to produce an individual whose genetic inheritance is neither that of the husband nor that of the wife who is "host"—I should say, "hostess"—during the parturition of the fetus; or, finally, parthenogenesis (scientifically induced virgin birth).

Thus, the ethical question or questions to be raised concern the morality of the ends or objectives to be adopted in any program of genetic control, the morality of "progressive" eugenics in comparison with the morality of "preventive" eugenics, the morality of parental selection in comparison with the morality of genetic surgery, and the morality of each of the specific means that are currently proposed and are apt to come into greater prominence in future discussion. Such are the considerations that have to be brought under scrutiny if anyone asks about the moral and religious implications of this science-based issue.

The Ethics Science Presupposes

First, however, it will be illuminating to ask: what is the ethics actually governing in proposals for genetic control, and among geneticists themselves?

In its account of the origin of the unique, unrepeatable individual person, the science of genetics seems to have resolved an ancient theological dispute. The human individual first comes into existence as a minute informational speck which resulted from the random combination of a great number of still more minute informational specks derived from the genetic pool which his parents passed on to him at the moment of impregnation. His subsequent prenatal and postnatal development may be described as a process of becoming what he already is from the moment he was conceived. There are a virtually unimaginable number of combinations of paternal and maternal genes that did not come to be when these particular genes were refused and be began to be. It is, therefore, virtually certain that no two individuals (with a single exception, to be mentioned in a moment) in the whole course of mankind's existence have ever had or will ever have the same genotype.[20] This is a form of "traducianism," i.e. the theory that the never-to-be-repeated individual human being (the "soul" is the religious word for him) was drawn forth from his parents at the time of conception.

Thus, science seems to have demonstrated what theology never could. Of course, prescientific notions are still believed by a great many people. Our law, for example (which seems always to be based on antique notions and to be in need of reform), takes the moment of birth to be the moment after which there is a "man alive" (for which the evidence is air in the lungs). After that, he is what he is becoming; and only then is "murder" possible as a crime. Where "abortion" is defined as a criminal offense in our legal system, this is a separate category of proscribed actions. Abortion is not legally prohibited as a species of murder, but because of the law's presumption that society has a stake in the prehuman material out of which the unique individual is born. Or again, "superstitious" people—Roman Catholics, for example—still may, and do, debate whether the unique, never-to-be-repeated human being begins with impregnation ("traducianism") or at a later moment when fetal life becomes independently animate, perhaps self-moving, in the womb ("creationism"). If "mediate" animation is the moment when the individual offspring first begins to be what he is to become, and launches on a course of becoming what he already is, then direct abortion after animation would be—morally, not legally —a species of murder; while direct abortion before animation, if this is defined as an offense, would fall within a class of far less serious invasions of the "nature" of our generative faculties, like committing 'adultery.'

Modern genetics, however, seems to have settled all this when it demonstrated, if not quite the unrepeatability, at least the never-to-be-repeated character, of that first informational speck each of us once was and still is in every cell and attribute.

I may pause here to raise the question whether a scientist has not an entirely "frivolous conscience" who, faced with the awesome technical possibility that soon human life may be created in the laboratory and then be either terminated or preserved in existence as an experiment, or, who gets up at scientific meetings and gathers to himself newspaper headlines by urging his colleagues to prepare for that scientific accomplishment by giving attention to the "ethical" questions it raises—if he is not at the same time, and in advance, prepared to stop the whole procedure should the "ethical finding" concerning this fact-situation turn out to be, for any serious conscience, murder. It would perhaps be better not to raise the ethical issues, than not to raise them in earnest.

This genetic account of the origin of human individuality discloses its need for supplementation (though, of course, not its incorrectness as one of the sciences of man) in the case of the single exception to this contemporary traducianism. This is the case of identical twins. Identical twins have the same genotype. They arise from the same informational speck. Yet each is, and knows he is, a unique, unrepeatable human person—something he is that he never was by virtue of his genotype; something he became, at some time and in some manner, for which the genes drawn from his parents do not account, or that he was not already from the fission following the original conception.

To explain the difference between identical twins one may resort to differences in the environment, which are great even in the normal case of identical twins who grow up in the same household. This would be a modern formulation of the theory of "creationism": the Environment, the maker of all twin-differences and the creator of a twin's unshareable individual being, "infused" this into the original hereditary material which was the same genotype. Thus, in the case of identical twins, a human person becomes himself from what he never was at the beginning or by genetic determination alone; and it does not matter that he is in process of becoming who he is for the duration of an entire lifetime in one environment or another. Also, it would not matter that one cannot tell exactly when, in the prenatal environment, killing an identical twin fetus is to kill not him but his brother, because each is, as it were, his own twin and because each has, so far, no more of his individual humanity than the identical genotype.

The question to be raised is whether two such sciences are more sufficient than one (though both are doubtless correct). Does the drawing forth of the person from his informational speck and from his environment give a complete account of man—his uniqueness, individuality, and unrepeatability? According to the total genetic-environmental vision of life, it would seem that, given infinite time, all possibilities would be actualized—including events that are so extremely unlikely as to be almost inconceivable. Then two individuals of the same genotype might arise from two different combinations of sexual determiners. These two individuals would be, as it were, "identical twins" even though they are born a hundred thousand years apart in time. Or, we may suppose two genotypes, known to be identical twins, brought up in separate but *identical* environments. These latter individuals would differ only in their *that-ness* and *where-ness,* as similar sticks and stones or atoms and neutrons may differ; but they would not be significantly distinct in their *what-ness* or *who-ness.* Thus, from some of its tendencies, modern thinking would accomplish a return to the vision of those ancients who believed "that the same periods and events of time are repeated [or are, at least, essentially repeatable]; as if, for example, the philosopher Plato, having taught in the school at Athens which is called the Academy, so, numberless ages before, at long but certain intervals, this same Plato and the same school, and the same disciples existed, and so also are to be repeated during the countless cycles that are yet to be.[21]

Now, I am sensible that there is a great distance from the conclusion of genetics and environmentalism that an identical human individual is *"almost* certain" not to be repeated, to the conclusion of the ancients that the repeatable *will* certainly be repeated.[22] Still from the idea that the individuals who appear in biological evolution have such a genetic constitution that it is *exceedingly unlikely* that any one of them will ever recur, an equally large leap is needed to arrive at the notion that in himself the individual has uniqueness of such a kind that he *cannot* be repeated or replayed.

I do not propose to develop here a counterthesis concerning the nature of man, or to try to demonstrate that theology is still the queen of the sciences. It is more important to point out that only if there is a more adequate view of man will ethics be possible, and that where there is ethics, there will be implied a more adequate view of man than the one we have just reviewed. The geneticists I have read are themselves instances of this. There is no ethics to be found among the contents of any science. There is, nevertheless, a morality of science. Geneticists, notably, do not treat the individual as if he were merely the carrier of the genetic determiners that will be productive of the next and future generations. They do not reduce him to the red color or the sweetness of a ripe apple fallen to the ground (of which it can perhaps be said that it is "almost certain" there

has never been another quite like it)—an apple which,
engorged by an animal who defecates the seed at a distance from the parent tree, secures the spread of the
species. Geneticists do not treat the individual as if he
were that, plus what the environment makes him. Few
are the geneticists who, in their proposals concerning
genetic control, toy with the idea that there might be a
chemical added to the water supply that would make
everyone sterile, and then a second chemical that would
reverse the effect of the first and would be given by the
government to selected persons "licensed to bear children."[23] The contemporary geneticist, Professor H. J.
Muller, who is most pessimistic about our "genetic
load" and its rapid deterioration under the conditions of
modern life and who is therefore most radical in his
proposals for genetic improvement, is also the most in-
sistent upon the use of voluntary means only, and upon
achieving his imperative goals by reliance on the exercise
of responsible human freedom.[24]

It is true that H. J. Muller sometimes writes as if he
utterly disparages the *present* individual human being.
"We thus arrive," he writes, "at a picture of humankind
as a creature living between an enormous past and perhaps a still longer future, fabricated out of inner micro-
somic immensities that he cannot directly perceive, but
on the surface of which, as it were, his awareness floats
as on a film, while facing outer vistas of a staggering
grandeur. He thus stands poised among a series of
abysses." Still the whole dignity of this awareness that
floats as on a film evidently consists in the fact that he
knows this, and can act accordingly, for Muller concludes the foregoing sentence with the words: "yet by
searching and studying these abysses, he is finding
means of traversing them."[25] Again, Muller writes that

"a more realistic consideration . . . will usually show
. . . that we have about as much to be ashamed of in
ourselves genetically as to be proud of. It will reveal us
as being one small even though potentially significant
mote in the whole human assemblage, a mote constitu-
tionally inclined to over-estimate itself."[26]

It is true that, as this passage shows, obsessed by the
gloomy facts about mankind's genetic load, Muller does
not accept (as Augustine said of the Manichees) with
good and simple faith this good and simple reason why
the good God created such a world as this—because, for
all the mutations corrupting it, its basic nature still is
good. It is also true that, like Engels, he believes in his
pessimism that everything that is real within the realm
of generation is bound to become unreasonable after a
while; hence it is already by definition unreasonable;
and in his optimism he believes that everything that is
reasonable within the hearts of men may become real,
however much it may contradict the existing seeming
genetic reality. Still, nothing in this excludes the governing influence of the ethics described above. The man of
the future will be both a product and a conscious agent
whose dignity is exhibited by his transcendent control
over his own evolution. And while now man is only a
mote among the contents of genetics, he is also a thinking mote who can rise above all this and intend the
world as a geneticist, and whose noblest action would be
to sacrifice his existing seeming reality to that future
ideal.

We shall have to ask what is the *source* and what is the
rational or other *ground* of the ethics that governs eugenic proposals. The source of this morality is doubtless
the atmospheric humanism and the liberal progressivism
which sustains any ostensibly science-based ethics to-

day. Anyone who reads the numerous, morally impassioned writings of H. J. Muller on the subject of eugenics—writings composed by a man who can almost hear the "sound" of deleterious mutations going on all around him and who knows these increasingly fail to be eliminated from the human genetic pool—is apt to exclaim that there has been nothing like it since Condorcet wrote his treatise on human progress in a Paris prison cell within the sound of the guillotine clicking away at its fell work amid the roars of a degenerate mob!

When one asks about the rational or other grounds of the ethics of voluntarism and responsible freedom (the humanitarism and progress doctrine) that evidently is governing in proposals for genetic improvement, and when one asks for an account of the respect for the uniqueness and dignity of the human individual, made manifest by and entailed in this ethic, no sufficient answer can be found in the science of genetics itself, or in the truth it discovers concerning the biological world. The answer must be found (and for science this is the final answer) among the presuppositions of there being any science of genetics at all, not in its contents. There is an ethic, and there is a view of man that makes science possible. Man must be, his mind must be, and his virtues or values must be of a certain order for there to be the preconditions of any scientific knowledge at all. And as Kant knew long ago, anything that is a *necessary* presupposition of scientific knowledge must be as certainly valid as that scientific knowledge itself.

"All biological species have evolved, but only man knows that he has evolved."[27] Biologically speaking, man is a creature with a certain favorable or unfavorable genetic inheritance; but, still biologically speaking, man is the geneticist who knows this. Genetically speaking, we may say à *la Pascal,* man is a thinking mutant. It is not necessary for the whole universe to arm itself to crush him. A vapour, a drop of water, a mutant gene, is sufficient to slay him. But were his genetic load to crush him, man would still be nobler than that which kills him, for he knows that he dies, he knows all the truths of microgenetics and he foreknows that he may die of the degenerative forces accumulating in the genetic substructure from which came also his genetic superiority over other forms of life, while the universe and the lethal genes know nothing of the advantage they have over him. Thus our whole dignity consists in thought.[28] Thus, the ethics of science applicable to this science-based issue is the fruit of consciously intending the world as a geneticist; and the view of man here entailed is that of man the truth-seeking, truth-finding and truth-using animal. Man is the free and intentional *subject* of all this. He is not merely the object discovered among the contents of the science of genetics. Rather he is essentially the one who knows all that, and who may use his knowledge for the good of other persons. There is a genuinely humanistic ethics at work in proposals for genetic control, which finds it contradictory to the very possibility of such scientific knowledge that one might propose to treat the human individual as a mere object of genetic or environmental determination which would be coersively imposed upon him.

J. Bronowski formulated in general terms the *pre-suppositional* status of a scientific ethic (which must be at least as valid as any of the material findings of science) in *Science and Human Values;* he undertook to show "the place of science in the canons of conduct which it has still to perfect." In fact, he showed the place of certain canons of conduct in the history and

culture of science. He demonstrated that these canons
of conduct are the *necessary* presupposition of there
being any place whereon science itself might stand. In
fact, as Bronowski wrote, he brought under study "the
[moral] conditions for the success of science and
[found] in them the values of man which science would
have had to invent afresh if man had not otherwise
known them."[29] Chief among these canons of conduct
is "the habit of truth" which has made our society as
surely as it has made the linotype machine or Darwin's
Origin of Species. Another is the knowing-community
that science presupposes, and the fact that verification
has no meaning if it is assumed to be carried out by one
man without any reference to this community of discourse.[30] Thus, truth depends on "truthfulness"—"a
principle which binds society together, because without
it the individual would be helpless to tell the true from
the false." Bronowski formulates a categorical imperative which is a necessary presupposition, or the condition, of the possibility of scientific knowledge: "We
ought to act in such a way that what *is* true can be
verified as such."[31]

The virtues and standards this requires, however much
scientists may sometimes fail to achieve them, are,
nevertheless, not an alien professional code sought to be
imposed. Instead, the virtues prerequisite to science
"spring from the pith and sap of the work they regulate"; the real essence of scientific work is brought
against any existential distortions of it. Thus, the values
of science derive neither from the personal superiority
of its participants nor from "finger-wagging codes";
"they have grown out of the practice of science, because
they are the inescapable conditions for its practice."
"An ethics for science . . . derives directly from its own

activity." This ethics includes the values of tolerance and democracy, since "science cannot survive without justice and honor and respect between man and man." Finally, by this presuppositional method of justification, Bronowski affirms the unique value of the individual: science "must prize the search above the discovery and the thinking (and with it the thinker) above the thought. In the society of scientists each man, by the process of exploring for the truth, has earned a dignity more profound than his doctrine. A true society is sustained by the sense of human dignity."[32] Thus again it is manifestly the case that, as Pascal said, our whole dignity consists in thought—no matter what the discovery, or the 'doctrine' (e.g. the science of genetics, objectively viewed), says about man.

This, then, is the ethics that governs most genetic proposals. It is an ethics that is rationally grounded in the fact that, genetically speaking, man is at least the geneticist—and all else that this fact implies.

Upon observing this to be the case, we may well find it significant that when Bronowski writes that science would have had to invent these values, if they did not already exist, he said: "if man had not otherwise *known* them." Such an ethics is the fruit of intending the world as a man in the community of men, and not simply the fruit of intending the world as a scientist in the community of scientists. Doubtless in the present age it is important that science confirms these human values, that it perfects them, and "in societies where these values do not exist, science has had to create them."[33] Still, this ethic must already be grounded in a more adequate understanding of what it means to be a man than is contained in, or can be forthcoming from, the "doctrine" of the individual's genetic origins (however correct this may be), or, for that matter, from "the habit of truth itself."[34] Western science is itself the product of a certain civilization and its values, which are 'otherwise known' if they are ever known to be valid; science may only perfect them and, in other cultures, help to create them.

Finally, it is pertinent to make this concluding observation—without prejudice, without forgetting the extraordinary richness of those values which can be shown to be the necessary presupposition of there being any science at all, and without disparaging the depth in the corollary that the individual's whole dignity consists in thought. A great deal is here asserted to be morally known that is not simply contained in the contents of any science. Still, this is a limiting view; and the limits arise from the fact that the ethics is a fruit of intending the world as a scientist, and not expressly from intending the world as a man among men (much less from intending the world as a Christian or as a Jew). This accounts for the fact that when geneticists begin to describe those human qualities to be selected and bred into the race of men, they write remarkably as if they were describing the attributes of mind and of character that would make a good geneticist, or at least a good community of scientists. Acknowledging that these are notable humanistic values, still there are other modes of being human.

The Genetic Apocalypse and the End of Man

In order to analyze the moral implications of genetic control for western religions, it is necessary to lift up to view certain aspects of what it means to intend the world as a Christian or as a Jew. These also are modes of

Ethics known by thought + by will

being human, and of how values are "otherwise known" in this world and ethical judgments made. On the assumption that it is a Christian *subject* who has come into the possession of all this genetic knowledge and who faces our genetic dilemma, what will be the attitude he takes toward eugenic proposals? Two ingredients are of chief importance. First, we have to contrast biblical or Christian eschatology with genetic eschatology, and observe how these practical proposals may change their hue when shifted from one ultimate philosophy of history to the other. This will be the matter of the present section of this paper. Then, secondly (in the following section), we have to explore the bearing which the Christian understanding of the union between the personally unitive purpose and the procreative purpose of human sexual relations (sex as at once an act of love and an act of procreation) may have upon the question of the means to be used in genetic control.

The writings of H. J. Muller give the most vivid portrayal of the genetic cul-de-sac into which the human race is heading. He describes, in fact, a genetic apocalypse. His fellow geneticists can correct, if they must, the extremism of this vision. For the purpose of making clear, however, how one intends the world as a Christian, even in the face of such an apocalyptic account of the end toward which we are proceeding, or which is coming upon us, it is better to leave the vision unaltered and assume it to be a true account of the scientific facts.

Within a period of a few million years, according to Muller, provided that during this period our medical men have been able to continue to work with the kind of perfection they desire, "the then existing germ cells of what were once human beings would be a lot of hopeless, utterly diverse genetic monstrosities." Long before that, "the job of ministering to infirmities would come to consume all the energy that society could muster," leaving no surplus for general or higher cultural purposes.[35] People's time and energy would be mainly spent in an effort "to live carefully, to spare and prop up their own feebleness, to soothe their inner disharmonies and, in general, to doctor themselves as effectively as possible." Everyone will be an invalid, and everyone's accumulated internal disability would amount to lethality if he had to live under primitive conditions.[36] If any breakdown occurs in the complex hospital system that civilization will have become, mankind will be thrown back into a wretchedness with which his primitive beginnings cannot be compared.

Our descendants' natural biological organization would in fact have disintegrated and have been replaced by complete disorder. Their only connection with mankind would then be the historical one that we ourselves had after all been their ancestors and sponsors, and the fact that their once-human material was still used for the purpose of converting it, artificially, into some semblance of man. However, it would in the end be far easier and more sensible to manufacture a complete man *de novo*, out of appropriately chosen raw materials, than to try to refashion into human form those pitiful relics which remained. For all of them would differ inordinately from one another, and each would present a whole series of most intricate research problems, before the treatments suitable for its own unique set of vagaries could be decided upon.[37]

It is unreasonable to expect medicine to keep up with the problem (especially because medical men themselves in that near, or distant, future will be subject to the same genetic decomposition); "at long last even the most sophisticated techniques available could no longer suffice to save men from their biological corruptions"[38] (and, again, I add to Muller's assumptions, medicine in that future could not be all that sophisticated, because of the genetic deterioration of the medical men who would be alive in the generation before the genetic *eschaton*).

Stripped of rhetoric, this means that, according to the genetic apocalypse, there shall come a time when *there will be none like us to come after us.* There have been other such scientific visions of the future. Whether this results from the pollution of our atmosphere and water by industrial refuse, or of the atmosphere by strontium 90, or from a collision of planets, the burning up of the earth, or the entropy of energy until our planet enters the eternal night of a universe run down, these scientific predictions—without exception—portray a planet no longer fit for human habitation, or a race of men no longer fit to live humanly. Because these are science-based apocalypses, the gruesome details of the "last days" can be filled in, and our imagination heightened in its apprehension of the truth concerning physical nature and the prospects of human history in the one dimension that is scientifically known to us. All these visions quite realistically teach that there will come a time when there will be none like us to come after us. It is as obvious as the ages are long that it is an infirm philosophy which teaches that "man can be courageous only so long as he knows he is survived by those who are like him, that [in *this* sense] he fulfils a role in something more permanent than himself."[39] Every scientific eschatology (with the single exception of the view that human history is eternal) places in jeopardy courage and all other values that are grounded in the future of the human generations. It does not matter whether the end comes early or late. Nor do the gruesome details do more than heighten the imagination. They do not add to the ultimate meaninglessness to which all human affairs were reduced when meaning came to rest in the temporal future (unless that future is foreknown to be eternal—and, if one thinks this through, it too is a melancholy prospect). All that can be said to the credit of the genetic apocalypse, or to the credit of any science-based eschatology, is that it makes *impressive* the truth that was already contained in the thought that men live in "one world."

Anyone who intends or perceives the world as a Christian will have to reply that he knew this all along, and that he has already taken into his system the idea that one day there will be none like us to come after us. Even gruesome details about what will happen in the "last days" are not missing from the Christian's Apocalypse, even though admittedly these are not extrapolations from scientific facts or laws. The Revelation of St. John is still in the Bible; and even the so-called little apocalypse (Mark 13 and parallels) had this to say: "In those days shall be affliction, such as was not from the beginning of the creation which God created unto this time, neither shall be. . . . But in those days, after that tribulation, the sun shall be darkened, and the moon shall not give her light, and the stars of heaven shall fall, and the powers that are in heaven shall be shaken" (Mark 13:19, 24-25). Again, stripped of rhetoric, there will be none like us to come after us on this planet.

This means that Christian hope into, and through, the future depends not at all on denying the number or seriousness of the accumulating lethal mutations which Muller finds to be the case (let his fellow geneticists argue with him however they will).

Where genetics teaches that we are made out of genes and unto genes return, Genesis teaches that we are made out of the dust of the ground and unto dust we and all our seed return. Never has biblical faith and hope depended on denying or refusing to face any facts—either of history, or of physical or biological nature. No natural or historical "theodicy" was ever required to establish the providence of God, for this providence was not confined to the one dimension within which modern thought finds its limits.

It is as easy (and as difficult) to believe in God after Auschwitz, as it was after Sennacherib came down like a wolf on the fold to besiege and destroy the people of God. The Jews who chanted as they went to meet their cremation, *"Ani Ma'amin . . . "*—"I believe with unswerving faith in the coming of the Messiah"—uttered words appropriate to that earlier occasion, and to all temporal occasions. It is as easy (and as difficult) to believe in God after Mendel and Muller, as it was after Darwin or the dust of Genesis. Religious people have never denied, indeed they affirm, that God means to kill us all in the end, and in the end He is going to succeed. Anyone who intends the world as a Jew or as a Christian—to the measure in which this is his mode of being in the world—goes forth to meet the collision of planets or the running down of suns, and he exists toward a future that may contain a genetic apocalypse with his eye fixed on another *eschaton: "Ani Ma'amin . . . "* He may take these words literally, or they may imaginatively express his conviction that men live in "two cities" and not in one only. In no case need he deny whatever account science may give him of this city, this history, or this world, so long as science does not presume to turn itself into a theology by blitzing him into believing that it knows the one and only apocalypse.[40]

This does not mean a policy of inaction, or mere negative acceptance, of trends in history or in biology on the part of anyone who is a Christian knowing-*subject* of all that he knows about the world. Divine determination, properly understood, imposes no iron law of necessity, no more than does genetic determination. Only the ultimate *interpretation* of all the action that is going on is different, and significantly different. We shall have to ask what practical difference this makes as one man goes about responding (in all the action that comes upon him) to the action of the laws of genetics, while another goes about responding (in all the action coming upon mankind) to the action of God; or, as one gives answers to the ultimate untrustworthiness of the force behind genetic trends, while another answers with his life and choices to a trustworthiness beyond all real or seeming untrustworthy things.[41]

The differences are two—one pervasive and the other precise. In the first instance, one must notice the tone of assertive or declaratory optimism based on the ultimate and unrelieved pessimism that pervades the thought of some proponents of eugenics. The writings of H. J. Muller cannot be accounted for simply by the science of genetics, or even by the fact that his ethics is that of a man who intends the world as a scientist and who finds the whole dignity of man to consist in thought. As such, and in themselves, these things might be productive of more serenity, or serenity in action. But it is the whole creation, as it is known in genetics to be effectively present today and into the future, that

Muller is fighting. No philosophy since Bertrand Russell's youthful essay[42] has been so self-consciously built upon the firm foundations of an unyielding despair. Mankind is doomed unless positive steps are taken to regulate our genetic endowment; and so horrendous is the genetic load that it often seems that Muller means to say that mankind is doomed no matter what steps are taken. Yet his optimism concerning the solutions he proposes is no less evident throughout; and all the more so, the more it is clear that his solutions (dependent as they are upon voluntary adoption) are unequal to the task. The author's language soars, he aspires higher, he challenges his contemporaries to nobler acts of genetic self-formation and improvement, all the more because of the abyss below. The abyss sets up such powerful wind currents that mankind seems destined to be drawn into it no matter how high we fly. These are some of the consequences of the fact that when all hope is gone, Muller hopes on *in despair.* An Abraham of genetic science, if one should arise, would be one who, when all hope is gone, hopes on *in faith,* and who therefore need neither fear the problem nor trust the solution of it too much.

The more precisely identifiable difference is the greater room there will be for an "ethics of means" in the outlook of anyone who is oriented upon the Christian *eschaton* and not upon the genetic cul-de-sac alone. Anyone who intends the world as a Christian or as a Jew knows along his pulses that he is not bound *to succeed* in preventing genetic deterioration, any more than he would be bound to retard entropy, or prevent planets from colliding with this earth or the sun from cooling. He is not under the necessity of *ensuring* that those who come after us will be like us, any more than he is bound to *ensure* that there will be those like us to come after us. He knows no such *absolute* command of nature or of nature's God. This does not mean that he will do nothing. But it does mean that as he goes about the urgent business of doing his duty in regard to future generations, he will not begin with the desired *end* and deduce his obligation exclusively from this end. He will not define *right* merely in terms of conduciveness to the good end; nor will he decide what *ought to be done* simply by calculating what actions are most likely to succeed in achieving the *absolutely imperative end* of genetic control or improvement.

The Christian knows no such absolutely imperative end that would justify any means. Therefore, as he goes about the urgent business of bringing his duty to people now alive more into line with his genetic duty to future generations, he will always have in mind the premise that there may be a number of things that might succeed better but would be intrinsically wrong means for him to adopt. Therefore, he has a larger place for an ethics of means that is not wholly dependent on the ends of action. He knows that there may be a great many actions that would be wrong to put forth in this world, no matter what good consequences are expected to follow from them—especially if these consequences are thought of simply along the line of temporal history where, according to the Christian, success is not promised mankind by either Scripture or sound reason. He will approach the question of genetic control with a category of "cruel and unusual means" that he is prohibited to employ, just as he knows there are "cruel and unusual punishments" that are not to be employed in the penal code. He will ask, What are right means? no less than he asks, What are the proper objectives? And he will know in advance that any person, or any society or age, expecting ultimate success where ultimate success is not to be reached, is peculiarly apt to devise extreme and morally illegitimate means for getting there. This, he will know, can easily be the case if men go about making themselves the lords and creators of the future race of men. He will say, of course, of any historical and future-facing action in which he is morally obliged to engage: "Only the end can justify the means" (as Dean Acheson once said of foreign policy). However, because he is not wholly engaged in future-facing action or oriented upon the future consequences with the entirety of his being, he will immediately add (as Acheson did): "This is not to say that the end justifies any means, or that some ends can justify anything."[43] An ethics of means not derived from, or dependent upon, the objectives of action is the immediate fruit of knowing that men have another end than the receding future contains.

The ethics which, as we have seen, governs genetic proposals says as much. A fruit of intending the world as a geneticist is an ethics whose means are determined by the values of free will and thought. This puts a considerable limit upon the actions which can be proposed for the prevention of the genetic apocalypse (which, if a correct prediction, belongs only to the *contents* of the science of genetics). Still, this is not a sufficient substance for the morality of action, or at least not all the substance a Christian will find to be valid. One who intends the world as a Christian will know man's dignity consists not only in thought or in his freedom, and he will find more elements in the nature of man which are deserving of respect and should be withheld from human handling or trespass. Specifically in connection with genetic proposals, he will know that there are more ways to violate man-womanhood than to violate the *freedom* of the parties; and that something voluntarily adopted can still be wrong. He will pay attention to this as he goes about using indifferent, permitted, or not immoral means to secure the *relatively* imperative ends of genetic control or improvement. To this ethics of means we turn in the next section.

The Covenant of Marriage and Right Means

In relation to genetic proposals, the most important element of Christian morality—and the most important ingredient that the Christian acknowledges to be deserving of respect in the nature of man—which needs to be brought into view is the teaching concerning *the union between* the two goods of human sexuality.

An act of sexual intercourse is at the same time an act of love and a procreative act. This does not mean that sexual intercourse always in fact nourishes love between the parties or always engenders a child. It simply means that it *tends, of its own nature,* toward the strengthening of love (the unitive or the communicative good), and toward the engendering of children (the procreative good). This will be the nature of human sexual relations, provided there is no obstruction to the realization of these natural ends (for example, infertility preventing procreation, or an infirm, infertile, or incurving heart that prevents the strengthening of the bonds of love).

Now, there has been much debate between Protestants and Roman Catholics concerning whether the unitive or the procreative good is primary, and concerning the hierarchial order or value-rank to be assigned these goods. I have shown elsewhere[44] that, contrary to popular belief, there is in the present day little or no essential difference between Roman Catholic and Protestant teachings on this point. The crucial question that remains is whether sexual intercourse as an act of love should ever be separated from sexual intercourse as a procreative act. This question remains to be decided, even if the unitive and procreative goods are equal in primacy, and even if it be said that the unitive end is the higher one. It still must be asked, Ought men and women ever to put entirely asunder what God joined together in the covenant of the generating generations of mankind? Assign the supreme importance to sexual intercourse as an act of personal love, and there still remains the question whether, in what sense, and in what manner, intercourse as an act of love should ever be divorced from sexual intercourse as in and of itself procreative.

Now, I will state as a premise of the following discussion that an ethics (whether proposed by nominal Christians or not) that *in principle* sunders these two goods—regarding procreation as an aspect of biological nature to be subjected merely to the requirements of *technical* control while saying that the unitive purpose is the free, human, personal end of the matter—pays disrespect to the nature of human parenthood. *Human* parenthood is not the same as that of the animals God gave Adam complete dominion over. Such a viewpoint falls out of the bounds which limit the variety of Christian positions that may be taken up by, and debated among, people who undertake to intend the world as Christians.

It is important that these outer limits be carefully defined so that we see clearly the requirements of respect for the created nature of man-womanhood, and so that we do not rule out certain actions that have traditionally been excluded. Most Protestants, and nowadays a great many Catholics, endorse contraceptive devices

which separate the sex *act* as an act of love from whatever tendency there may be in the act (at the time of
the act, and in the sexual powers of the parties) toward
the engendering of a child. But they do *not* separate the
sphere or realm of their personal love from the sphere or
realm of their procreation, nor do they distinguish between *the person* with whom the bond of love is nour-
ished and *the person* with whom procreation may be
brought into exercise. One has only to distinguish what
is done in particular *acts* from what is intended, and
done, in a whole series of acts of conjugal intercourse in
order to see clearly that contraception need not be a
radical attack upon what God joined together in the
creation of man-womanhood. Where planned parenthood is not planned *un*parenthood, the husband and
wife clearly do not tear their own one-flesh unity completely away from all positive response and obedience to
the mystery of procreation—a power by which at a later
time their own union originates the one flesh of a child.

Moreover, the fact that God joined together love and
progeny (or the unitive and procreative purposes of sex
and marriage) is held in honor, and, not torn asunder,
even when a couple for grave, or for what in their case is
to them sufficient, reason adopt a lifelong policy of
planned *un*parenthood. This possibility can no more be
excluded by Protestant ethics than it is by Roman Catholic ethics, which teaches that under certain circumstances a couple may adopt a systematic and possibly
lifelong policy of restricting their use of the unitive
good to only such times as, it is believed, there is no
tendency in the woman's sexual nature toward conception. The "grave reasons" permitting, or obliging this,
have been extended in recent years (the original "reason" being extreme danger to a woman's life in child-

birth) to include grave family financial difficulties (because the end is the procreation *and education* of the child). These "grave reasons" have been extended even to the point of allowing that the economy of the environing society and world overpopulation may be taken into account, by even the healthy and the wealthy, as sufficient reason for having fewer children or for having no more at all.[45] Once mankind's genetic dilemma is called to the attention of the Church and its moral theologians, I see no intrinsic reason why these categories of analysis may not be applied to allow ample room for voluntary eugenic decisions, either to have no children or to have fewer children, for the sake of future generations.

After all, Christian teachings have always held that by procreation one must perform his duty to the future generations of men; procreation has not been a matter of the selfish gratification of would-be parents. If the fact-situation disclosed by the science of genetics can prove that a given person cannot be the progenitor of healthy individuals (or at least not unduly defective individuals) in the next generations, then such a person's "right to have children" becomes his duty not to do so, or to have fewer children than he might want (since he never had any right to have children simply for his own sake). Protestant and Roman Catholic couples in practicing eugenic control over their own reproduction may (unless the latter's church changes its teaching about contraception in the wake of the Vatican Council) continue to say to one another: you in your way, and I in God's! In the turmoil over Pope Paul VI's encyclical *Humanae Vitae,* one Catholic couple says, in effect, to another: you in your way, I in the Church's or at least the Pope's. Still, it is clear that the Roman Catholic no

less than the Protestant Christian could adopt a policy
of lifelong nonparenthood, or less parentage, for eugenic
reasons. Such married partners would still be saying by
their actions that if either has a child, or if either has
more children, this will be from their own one-flesh
unity and not apart from it. Their response to what God
joined together, and to the claim He placed upon human
life when He ordained that procreation come from sexual love, would be expressed by their resolve to hold
acts of procreation (even the procreation they have not,
or have no more) within the sphere of acts of conjugal
love, within the covenant of marriage.

Before going on to explore the implications of Christian ethics for methods of genetic control other than
voluntary, eugenically directed birth control, it may be
important to correct a misinterpretation of the principle
we are using (the union of conjugal love with procreation) and to show how cardinal a principle this is in
theological ethics and in the way Christians understand
the entire realm with which genetics deals. It might be
supposed that the moral judgments defining the outer
limits of responsible human conduct are based on a
mere fact of biological life, on the "natural law" in this
sense, on Genesis, or—as theologians would say—on the
first article of the Creed which speaks of "creation" and
the Creator. It is true that a Christian will refuse to
place man's own sexual nature in the class of the animals over which Adam was given unlimited dominion.
He will regard man as the body of his soul as well as the
soul of his body, and he is not apt to locate the *humanum* of man in thought or freedom alone. He will
also discern immediately that many prevalent modern
views are based on other "myths of creation." Many a
modern outlook would elevate personal freedom in the

expression of love without in any way honoring its connection with procreation. Instead these viewpoints call upon men and women to act as if anything that technically can be done to exert dominion over procreation may, or should, be done if it is voluntary and desirable in terms of consequences. Faced with these proposals, a Christian is apt to sum them all up in a "myth of creation" that is not his, namely, by rewriting Genesis to tell of the creation of man-womanhood with two separate faculties: sex serves the single end of manifesting and deepening the unity of life between the partners, while human offspring are born from the woman's brow and somehow impregnated through the ear by a cool, deliberate act of man's rational will.

Still, all this would be a misunderstanding of the honor and obedience the creature should render to his Creator, and of the source of a Christian's knowledge of this duty. It arises not from slavish obeisance to a fact of nature. The Prologue of John's Gospel (not Genesis) is the Christian story of creation which provides the source and standard for responsible procreation, even as Ephesians 5 contains the ultimate reference for the meaning and nature of conjugal love and the standard governing covenants of marriage. Since these two passages point to one and the same Lord—the lord who presides over procreation as well as the lord of all marital covenants—the two aspects of human sexuality belong together. The two aspects belong together not simply because they are found together in human and other animal bisexual reproduction.

It was out of His love that God created the entire world of His creatures. The selfsame love which in Ephesians 5 becomes the measure of how husbands should love their wives was, according to the Prologue,

with God before all creation, and without Jesus Christ
was not anything made that was made. Of course, we
cannot see into the mystery of how God's love created
the world. No more can we completely subdue the mystery (which is but a reflection of this) contained in the
fact that human acts of love are also procreative. Nor
can we know why this was made to be so (in contrast to
the more "rational" myth we constructed a moment
ago). Nevertheless, we procreate new beings like ourselves in the midst of our love for one another, and in
this there is a trace of the original mystery by which
God created the world because of His love. God created
nothing apart from His love; and without the divine love
was not anything made that was made. Neither should
there be among men and women (whose man-womanhood—and not their minds or wills only—is in the image
of God) any love set out of the context of responsibility
for procreation, any begetting apart from the sphere of
love. A reflection of God's love, binding himself to the
world and the world to himself, is found in the claim He
placed upon men and women in their creation when He
bound the nurturing of marital love and procreation together in the nature of human sexuality. Thus, the
Christian understanding of life stems from the second
article of the Creed, not from the first or from facts of
nature; and this is the source of the Christian knowledge
that men and women should not put radically asunder
what God joined together in creation. Thus a Christian,
as such, intends the world as God intends the world. We
men know this at the very center of the Christian faith
and revelation, and here "right" and righteousness are
defined in terms of aligning our wills with His.

Men and women are created in covenant, to covenant,
and for covenant. Creation is *toward* the love of

Christ.[46] Christians, therefore, will not readily admit that the energies of sex, for example, have any other primary *telos,* another final end, than Jesus Christ. Rather, they will find in the strength of human sexual passion (beyond the obvious needs of procreation) an evident *telos* of acts of sexual love toward making real the meaning of man-womanhood, nurturing covenant-love between the parties, fostering their care for one another, prefiguring Christ's love for the Church—whatever other substrata of purposes sexual energy may have that can be discovered by intending the world as a biologist. And in human procreativity out of the depths of human sexual love is prefigured God's own act of creation out of the profound mystery of his love revealed in Christ. To put radically asunder what God joined together in parenthood when He made love procreative, to procreate from beyond the sphere of love (AID, for example, or making human life in a test-tube), or to posit acts of sexual love beyond the sphere of responsible procreation (by definition, marriage), means a refusal of the image of God's creation in our own.

A science-based culture, such as the present one, of necessity erodes and makes nonsense out of all sorts of bonds and connections which a Christian sees to be the case. Thus, because of its atomistic individualism, modern thought delivers the verdict "guilty of monopoly" upon the definition of marriage as a mutual and exclusive exchange of right to acts which tend to the nurturing of love or unity of life, and to the engendering of children—when all that was meant by this definition is that there is a covenantal bond of life with life. And among geneticists, H. J. Muller, for one, delivers the verdict guilty of "genetic proprietorship that so many men hold dear" or "fixation on the attempted perpetu-

ation of just his own particular genes,"[47] and "feelings
of proprietary rights and prerogatives about one's own
germinal material, supported by misplaced ego-
tism"[48]—when all that is at stake in the historic ethics
of the western world, and actually in the minds of a
great many people today, is the bond to be held in respect between personal love and procreation. As explained above, what is at stake is about as far from
selfish proprietorship as anything can be—as far as marriage is from monopoly. There may be an irrepressible
conflict between the values governing in some genetic
proposals and the historic values expressed by Christians, but there is no reason for the conflict to be an
irrational one, or irrationally conceived. The irrational
element enters the conflict wherever there is evidently
such an unparalleled breakdown of our moral tradition
that men of science cannot even understand what is
being said in Christian ethical judgments. The verdicts
"monopoly" and "proprietorship over germinal material" become judgments upon a whole culture that produces great intelligences capable of uttering these verdicts, or incapable of understanding Christian ethics
except in terms of these absurdities.[49]

In the preceding paragraphs I have attempted to explain the substance of that "ethics of means" which
Christianity adds to scientific humanism's insistence
upon the use of only voluntary means in any program of
genetic control. I have tried to express what is morally
at stake for the Christian religious ethics, and the rationale it lays down for determining the nature and limits
of specifiably legitimate conduct in this area. We have
now to resume our examination of the various methods
that have been proposed for the control or improvement
of man's genetic inheritance, evaluating these in the

light of the requirement that there be no complete, or radical, or "in principle" separation between the personally unitive and the procreative aspects of human sexual life. By this standard there would seem to be no objection to eugenically motivated birth control, if the facts are sufficient to show that genetic defects belong among those grave reasons that may warrant the systematic, even lifelong, prevention of conception. A husband and wife who decide to practice birth control for eugenic reasons are still resolved to hold acts of procreation (even the procreation they have not, or have no more) within the sphere of conjugal love.

This understanding of the moral limits upon methods that may properly be adopted in voluntary genetic control leads, I would argue, to the permissibility of artificial conception control (no less than to the permissibility of the so-called rhythm method) and to the endorsement of voluntary sterilization for eugenic reasons. I know that many of my fellow Christians do not agree with these conclusions. Yet it seems clear that both are open for choice as means (if the ends are important enough)—provided Christian ethics is no longer restricted to the analysis of individual *acts* and is concerned instead with the coincidence of the *spheres* of personal sexual love and of procreation (the spheres to which particular actions belong). Neither the husband (or wife) who practices artificial birth control nor the husband who decides to have a vasectomy is saying by the total course of his life anything other than that *if* either marriage partner has a child, or more children, it will be within their marriage-covenant, from their own one-flesh unity and not apart from it. In principle, they hold together, they do not put completely asunder, what God joined together—the sphere of procreation, even the procreation they have not or have no more, and the sphere in which they exchange acts that nurture their unity of life with one another. They honor the union between love and creation at the heart of God's act toward the world of his creatures, and they honor the image of this union in the union of love with procreativity in their own man-womanhood. Their morality is not oriented upon only the genetic consequences which are believed to justify any voluntary means; nor is it only an ethic of inner intention which is believed to make any sort of conduct right. They *do* something, and are constantly engaged in doing it. This gives their behavior a character that is derived neither wholly from the desired results nor from subjective intention. Through a whole course of life they actually unite their loving and their procreativity (which, incidental to this, they have not). So they do not do wrong. They do no wrong that good may come of it. They do right that good may come of it. [In this moral reasoning, the present writer can see no difference between the case for contraception and the case for voluntary contraceptive sterilization, except in not unimportant differences in the findings of fact that may warrant the one form of birth control or the other—and except for the fact that as yet sterilization is ordinarily irreversible. Even in terms of the more static formulations of the past, it should certainly be said that a vasectomy may be a far less serious invasion of nature than massive assault upon the woman's generative organism by means of contraceptive pills.]

[I am aware that many Christians will not agree with these conclusions—even some who are not Roman Catholics will be in disagreement. I ask the latter simply to consider the following possible development in the basic structure of their own ethical analysis of these problems. [Suppose that in the near future Roman Catholic teachings effect the shift from act-analysis (from concentrating upon features of *the act* of conjugal intercourse that ought not to be put asunder) to concentration upon these same features within the order of marriage, the sphere or realm of marriage, that ought not to be radically separated. This, I predict, will be the theological-ethical grounds for any approval of the use of contraceptive devices in acts or in a series of acts of conjugal intercourse, if Catholic teachings make this advance. Approval of contraceptive devices will not be by reference to the indirect effects of the pill in regularizing the woman's menstrual cycle: if only the pill is approved, the continued sway of act-analysis upon Catholic teachings will be evident. But if the order of marriage (with the goods that should be held together *between the parties*) comes to the fore in Catholic moral reasoning, artificial contraception will be warranted generally, where there is sufficient reason for controlling reproduction.[50]

If this is the step taken, it seems to me impossible to avoid the conclusion that voluntary male sterilization (when this is ordinarily a reversible operation) will find a place among the means of contraception. Perhaps it will even be preferable to other means that might be chosen. Then, if there are reasons for the systematic and lifelong practice of birth control (already a conclusion reached by Catholic moral theology), and if serious genetic defect finds a place among the reasons grave enough to warrant having no children at all, or no more children, then vasectomy would seem to be in principle permissible, perhaps commendable, maybe morally obligatory. Finally, where there is sufficiently grave reason for systematic, lifelong birth control, Christian moral reflection need not wait on the assured *reversibility* of vasectomy in order to accept it as a means of birth control. All this follows in the wake of taking quite seriously what I have tried to suggest by saying that a man and a women do not set creation asunder or disobey their Creator's will when they honor the union of their love with their procreativity, even the procreativity they have not, or have no more, within the covenant of marriage, which is a cause between them that is greater than they. They do not procreate from beyond their marriage, or exercise love's one-flesh unity elsewhere.

The notation to be made concerning genetic surgery, or the introduction of some anti-mutagent chemical intermediary, which will eliminate a genetic defect before it can be passed on through reproduction, is simple. Should the practice of such medical genetics become feasible at some time in the future, it will raise no moral questions at all—or at least none that are not already present in the practice of medicine generally. Morally, genetic medicine enabling a man and a woman to engender a child without some defective gene they carry would seem to be as permissible as treatment to cure infertility when one of the partners bears this defect. Any significant difference arises from the vastly greater complexity of the practice of genetic surgery and the seriousness of the consequences if, because of insufficient knowledge, an error is made. The cautionary word to be applied here is simply the moral warning against culpable ignorance. The science of genetics (and medical practice based on it) would be obliged both to be fully informed of the facts and to have a reasonable and well-examined expectation of doing more good than harm by eliminating the genetic defect in question. The seriousness of this consideration arises from the serious matter with which genetic surgery will be dealing. Still, the culpability of actions performed in unjustifiable ignorance cannot be invoked as a caution without allowing, at the same time, that in the practice of genetic medicine there doubtless will be errors made in inculpable ignorance. But genetic injuries of this order would be *tragic*, like birth injuries under certain circumstances. They would not entail *wrong*-doing; nor should applications of genetic science be stopped until all such eventualities are impossible. That would be an impossible demand, which no morality imposes. (But see pp. 100-01, 116-20, below.)

The paradox is that the most unquestionably moral means of genetic control (direct medical action for the sake of the genotype by some "surgical" or chemical anti-mutagent before the genotype is produced) is technically the most difficult and distant in the future,[51] while a number of the means presently available (phenotypic breeding in or breeding out) are of quite questionable morality, and questionable for reasons that the voluntariness of the practice would not remove.

In the foregoing, an affirmative notation has been placed beside genetically motivated conception control and voluntary sterilization. Before going on to other methods of achieving "empirical," genotypic or parental selection (beside which a negative notation must be placed), let a moralist confess himself to be in a quandary after reading some of the scientific literature concerning the paradox just mentioned. Some geneticists stress the great strides that could be taken toward solving mankind's genetic dilemma if science achieves the competence to perform genetic surgery and to *direct* mutation, or back mutation. These point out how little could be accomplished by empirical parent selection, short of forcing the gene pool of the future through a very narrow corridor and, by compulsion, bringing about the genetic extinction of a great number of potential reproducers.[52] Other geneticists stress what can be achieved in negative or positive eugenics by the voluntary use of the means of germinal selection presently available. These point out how almost unimaginably difficult and distant—and by comparison, roundabout and unnecessary—is the perfection of genetic surgery. Here is a conflict of scientific judgments, and one which may entail a subtle and suppressed moral judgment among geneticists themselves resulting in the difference in their readings of the fact-situation and in prognosis. In any case, it is quite impossible for a moralist who is a nonscientist to make his way to an analysis that he is confident is soundly based. This may give him the freedom to reach moral conclusions in his own right. Nevertheless, a layman cannot know which opinion to endorse, nor whether there is an emerging consensus among geneticists, nor whether the disagreement is wholly scientific or partly moral, when confronted by the following, opposing statements:

(1) "The technology of human genetics is pitifully clumsy, even by the standards of practical agriculture. Surely within a few generations we can expect to learn tricks of immeasurable advantage. *Why bother now with somatic selection, so slow in its impact?*"[53]

(2) "It is preposterous to suppose that, in the foreseeable future, knowledge would be precise enough to enable us to say what substitution to make in order to effect a given, desired phenotypic alteration. . . . But to suppose that, after it had become possible, men would still be bound by the reproductive traditions of today,

preferring this ultra-sophisticated method of improvement to the readily available one of selecting donor material free from the given defect or already possessing the desired innovation—that would be a calumny on the rationality of the human race. It would be like supposing that in some technically advanced society elaborate superhighways were constructed to carry vehicles *on enormous detours* to avoid defiling hallowed domains reserved in perpetuity for their millions of sacred cows."[54]

I must say that this last quotation from H. J. Muller indicates that he may be sorely in need of instruction in the difference between sacred cows and that "sacredness in the temporal order" who is *man*. Muller, of course, respects man's quality as a thinking animal; he would not violate his freedom, and he challenges men to noble action. This ethics, we have pointed out, is not to be found among the contents of the science of genetics, but is rather the necessary presupposition of man the geneticist and the fruit of intending the world with a scientific mind. Or, perhaps, Muller's humanism is a fruit of intending the world as a man within the community of men. Christian ethics, too, is not found among the contents of any natural science, nor can it be disproven by any of the facts that such sciences know. It is a fruit of intending the world as a Christian. (There is no conflict here between religion and science, but a conflict between two philosophies.) The Christian understands the *humanum* of man to include the body of his soul no less than the soul (mind) of his body. In particular, he holds in honor the union of the realm or personal love with the realm of procreativity in man-womanhood, which is the image of God's creation in the midst of His love. Since artificial insemination by means of semen

from a nonhusband donor (AID) puts completely asunder what God joined together, this proposed method of genetic control or genetic improvement must be defined as an exercise of illicit dominion over man no less than the forcing of his free will would be. Not all dominion over man's own physical nature, of course, is wrong, but *this* would be—for the reasons stated above.

In outline, Muller proposes that "germinal choice" be secured by giving eugenic direction to AID (Julian Huxley called this technique "pre-adoption"—it has already become a minority "institution" in our society). Muller also proposes that comparable techniques be developed and employed: "foster pregnancy" (artificial inovulation) and parthenogenesis (or stimulated asexual reproduction). The enormous difficulties in the way of perfecting punctiform genetic surgery or mutational direction by chemical intermediaries impel Muller to concentrate on presently available techniques of parental selection. Similarly, the apparently small gains for the race that can be secured by negative eugenics (because the genes will continue in great numbers as recessives in heterozygotes) impel him to advocate positive or progressive eugenics.[55] In positive genetics, one does not have to identify the genetic defects, or know that they do not add vigor in hybrids. One has only to identify the desired genotype (no small problem in itself!) and breed for it.

Muller rejects the following practices: choosing a donor who is likely to engender a child resembling the "adopting" father, using medical students (notoriously not of the highest intelligence) or barhops, using AID only when the husband is infertile or the carrier of grave genetic defect, and keeping the matter secret. Instead he proposes the selection of donors of the highest proven physical, mental, emotional, and moral traits, and he suggests that publicity be given to the practice so that more and more people may follow our genetic leaders by voluntarily deciding to bestow upon their "children" the very best genetic inheritance instead of their own precious genes.

In order for this practice to be most effective, Muller proposes that a system of deep-frozen semen banks be established and that records of phenotypes be kept and evaluated. At least twenty years should be allowed to elapse before the frozen semen is used, so that a sound judgment can be made upon the donor's capacities. The men who earn enduring esteem can thus be "manifold-ed" and "called upon to reappear age after age until the population in general has caught up with them."[56] It is an insufficient answer to this proposal to point out that in his 1935 book, *Out of the Night,* Muller stated that no intelligent and morally sensitive woman would refuse to bear a child of Lenin, while in later versions Lenin is omitted and Einstein, Pasteur, Descartes, Leonardo, and Lincoln are nominated.[57] Muller might well reply either by defending Lenin or by saying that not enough time had elapsed for a sound judgment to be made on him.

To his fellow geneticists can be left the task of stating and demonstrating scientific and other socio-psycho-logical objections, some of which follow: (1) the genes of a supposedly superior male may contain injurious recessives which by artificial insemination would become widespread throughout the population instead of remaining in small proportion, as they now do;[58] (2) the validity of this proposal is not demonstrated by the present-day children of geniuses; (3) "it might turn out that parents who looked forward eagerly to having a Horowitz in the family would discover later that it was not so fine as they expected because he might have a temperament incompatible with that of a normal family," and "it is bad enough if we take responsibility only for the environment of our children; if we take responsibility for their genetic make-up, too, the guilt may become unbearable;"[59] (4) we know nothing about the mutation rate that would continue in the frozen germ cells; (5) the IQ's of criminals would be raised;[60] (6) we could not have a "healthy society" because not many men would be "emotionally satisfied by children not their own."[61] Without debating these issues, my verdict upon the eugenic use of semen banks has been negative, in terms of the morality of means which Christian ethics must use as its standard of judgment.[62]

However, a word more should be said. No disciplined analysis of the moral life should fail to say that, among wrong actions, some are wronger than others. For the Roman Catholic, for example, abortion is worse than artificial conception control. Among invasions of man's generative faculties, some are more serious than others. While Muller's proposals would constitute a very serious invasion, and an utter separation of the realm of procreation from the realm of conjugal love, it might be that, after serious reflection upon the genetic problem, a Christian moralist could reach the conclusion that the genetic motivation and probable consequences of Muller's AID would add to it a redeeming feature—but not that this feature is sufficient to place the practice in the class of *morally permitted* actions.

Moreover, the judgment that AID for genetic or any other purposes is morally wrong does not entail the conclusion that it should be prohibited by law. Not all sin should be defined as crime. Not all immoral conduct is a fit subject for prohibitive legislation, but only acts that seriously affect the common good. It can be debated whether or not one of a number of current proposals concerning AID touches the common good so deeply that it belongs in the class of those immoral practices which should also be declared illegal. It is true that AID touches the moral nature of human parenthood (and tries to define it in terms of what it is not) just as deeply as—Roman Catholics believe—legal divorce touches the nature of marriage (and attempts to define it contrary to its nature). Still, Roman Catholics do not always and everywhere teach that under no circumstances should there be legislation permitting and regulating divorce (which for them is morally impossible). When this con-clusion is reached, it is by making a distinction between "sin" and "crime," or between conduct which is or is not a fit subject for prohibitive legislation—legislation which must be ever watchful to mold the common good out of the actual ethos of the people whose affairs it regulates. AID is in an area in which Anglo-American law fairly bristles with contradictions which will soon have to be cleared up by prohibitory, or permissive and regulatory legislation, or by case law. I am suggesting that it may be possible to justify the legal enactment of AID without basing this on its moral justification.[63] If so—or if, in any case, this is the cumulative judgment our society is making—then I suppose that a trial can, and will, be made to see what can be accomplished eu-genically by education and action in accord with Mul-ler's proposals.

Finally, it ought to be pointed out that the practice of freezing and storing sperm cells has a possibly desirable connection with genetically motivated voluntary sterili-zation. As a complement of vasectomy, this practice would provide germinal insurance. Such insurance may

have some role to play (unless and until vasectomy becomes ordinarily a reversible operation) in encouraging
men with moderately serious genetic defects to limit
their offspring. Germinal insurance would fall within the
genetic decisions of persons who, in adopting voluntary
contraceptive sterilization for eugenic reasons, are still
resolved to hold together sexual love and procreation in
the sexual intercourse they have with one another.

Ends in View

Finally, we need to bring under scrutiny the ends or
objectives of genetic control, and the choice to be made
between negative eugenics (by breeding-out or by back-
mutation) and progressive eugenics (by breeding-in or
by the positive direction of mutation).

H. J. Muller had supreme confidence that those pio-
neering spirits who lead the way in this generation in the
employment of germinal selection can be trusted to
choose, from a variety of choice-worthy genotypes (described to them by the keepers of the semen banks), the
types that will be good for mankind to produce in greater numbers. "Can these critics," he asks, "really believe
that the persons of unusual moral courage, progressive
spirit, and eagerness to serve mankind, who will pioneer
in germinal choice, and likewise those who in a more
enlightened age will follow in the path thus laid down,
will fail to recognize the fundamental human values?"[64]
Muller expresses the guiding aims of particular eugenic
decisions in quite general terms: "practically all peo-
ples," he writes, "venerate creativity, wisdom, brotherli-
ness, loving-kindness, perceptivity, expressivity, joy of
life, fortitude, vigor, longevity."[65] Or again: "What is
meant by superior is whatever is conducive to greater

wisdom, cooperativeness, happiness, harmony of nature, or richness of potentialities."[66] This understanding of the goals of eugenic decisions may be open to the objection that, in animal husbandry, one has to have very narrowly defined criteria governing the selection to be made. It is less open to the objection stated by Theodosius Dobzhansky: "Muller's implied assumption that there is, or can be, *the* ideal human genotype which it is desirable to bestow upon everybody is not only unappealing but almost certainly wrong—it is human diversity that acted as a leaven of creative effort in the past and will so act in the future."[67] There is range enough, it would seem, in Muller's description of ideal man to permit a great variety of specific genotypes. The fact is that within these very general value assumptions, Muller counts on individual couples to pick the specific genotype they want to bestow on their "preadoptive" children.[68] "Couples so enlightened as to resort in this and the next generation to germinal choice will not require a corps of axiologists or sociologists to tell them what are the most crying genetic needs of the man of today."[69] Thus, Muller is confident that a host of particular choices, made by people who have concrete options, can be laid, as it were, end to end with similar choices made by people in succeeding generations. The choices of the latter people will doubtless improve as their genetic inheritance improves, thus producing a continuity of choices in the ascending direction of genetic improvement (which was formerly the work of natural selection). This hope is only exceeded by Muller's certainty that, unless man assumes the direction of his genetic goals, the descent of the species is the sole alternative expectation.

Place beside this the objection, raised by Donald M.

Mackay, based on the fact that the generation that first initiates genetic control cannot determine the goals that will be set by future generations—or establish any directional continuity. No one can prevent "the 'goal-setting' from drifting and oscillating as time goes on, under the influence of external or even internal factors." Suppose genotype X is chosen in a majority of instances in the first generation. No one can know "what kind of changes these men of type X would think desirable in their successors—and so on, into the future." If we cannot answer this question and establish a continuity from the beginning, then "to initiate such a process might show the reverse of responsibility, on any explication of the term." (Moreover, unless this question is answered and unless future answers to it are assured, then the process would be quite unlike animal husbandry.) "In short, to navigate by a landmark tied to your own ship's head is ultimately impossible."[70]

Now, how does one adjudicate between these opposing views? It is obvious that these judgments fall far outside the science of genetics itself. There may even be operative here a kind of ultimate determination of one man's individual mode of being in the world toward making man the creator and determiner of his own evolution, and, on the part of the other scientist, a personal determination away from that dizzy prospect. The present writer would say that in order to refuse to concede some degree of truth to Muller's opinion, one has to be a rather thoroughgoing relativist who denies to man any fundamental competence to make moral judgments. This is why, in addition to supporting genetically motivated conception control and voluntary sterilization, I have conceded that, if AID is not to be prohibited by law, it might morally be a better wrong to use AID with the intention of bestowing a better genetic inheritance upon a child than if it is used with complete anonymity in regard to the donor's genotypic qualities and only for the sake of securing a child as much like the putative parents as possible. In any case, the voluntariness of the genetic decisions made in any one generation, and through the generations, insures the *usus* of Muller's proposal against such *abusus* as would forbid it *from the point of view of the ends only,* and would seem to render somewhat inconsequential such oscillation in goal-setting as might take place. Such oscillation in genetic decisions would be roughly comparable to oscillations in cultural decisions (taking place under the guidance of Muller's *jus gentium*) that may occur over the sweep of centuries; and the one would be no more and no less consequential than the other, while reciprocating strength to the other.

On the other side of this question, it must be acknowledged that this way of characterizing the goals to be set for positive human betterment does, despite its generality, describe the characteristics of a good geneticist, or the virtues of a good community of scientists, or at least the special values of man in the contemporary period. This is a science-based age, and an age of rapid social change in which men dream of inhabiting other planets after despoiling this one. It is an age in which "progressives" are in the saddle and ride mankind—ahead if not forward. In such an age it is natural enough that most of man's problems are defined in terms of "social lag" of one sort or another, and in terms of the laggard type of characters our genes continue to produce. Still, in the long view, mankind might be in the greatest peril if it succeeded in finding a way to increase its own momentum by selecting on a large scale for the special values of this present culture. In the long view, the race may have need of laggard types and traditional societies, who could take up the history of humanity again after the breakdown of the more momentous civilizations. If positive genetics gained its way, even under the aegis of a quite unexceptionable *jus gentium* setting the goals, would this not unavoidably take the form of genetically instituting some parochial *jus civilis?*

Partly because of the difficulties concerning goal-setting and because the negative goals would seem to be clearer, the present writer leans in the direction of approving preventive eugenics only. I do so also because the means to the ends of preventive genetics—whether these be the voluntary control of conception, or anti-mutagent surgery or chemicals—seem, at our present state of knowledge, to have the good effect of eliminating bad effects without as much danger of also producing an overflow of incalculable, unintended bad consequences. We may say with Hampton L. Carlson, "let us recommend preventive eugenics but proceed very cautiously in progressive eugenics. A firm scientific basis for the latter does not now exist."[71]

It must be admitted that the results on the total population of negative genetics may not be very effective in bringing about large-scale prevention of the deterioration of the gene pool. Nevertheless, faced by such pessimistic predictions, "it is well to remember that every defective individual that can be avoided represents a positive gain."[72] Also, if genetics can identify the *carriers* of genetic defects and thus we no longer need restrict preventive genetics to persons who are identifiably unfit themselves—if a qualitative control of reproduction can wisely be adopted by, and at some time in the future back-mutation can be performed helpfully upon, a larger proportion of the population—then the results of preventive eugenics need not be so limited as it has been in the past. To sterilize *forcibly* all persons suffering from serious genetic defect would have hardly any influence on the proportion of that particular recessive gene in the population. But if carriers can be identified, and if each heterozygous carrier has only half as many children as he would otherwise have, we would reduce the abnormal-gene frequency by fifty percent. This alone would greatly reduce the incidence of the defect in the next generation, and prevent untold future human misery.[73]

To make preventive eugenics more effective will require the development and widespread adoption of an "ethics of genetic duty." It is shocking to learn from the heredity clinics that have been established in recent years in more than a dozen cities in the United States, how many parents will accept grave risk of having defective children rather than remain childless. "When a husband and wife each carry a recessive deleterious gene similar to the one carried by the other, the chances of their having a defective child are one in four, with two children carriers of a single gene, but themselves without defect, and [only] the fourth child being neither a carrier nor defective. Couples in such a position, knowing that they have one chance in four of having a seriously defective child, and that two out of four of their children are likely to be carriers, still frequently take a chance that things will turn out all right."[74] This can only be called genetic imprudence, with the further notation that imprudence is gravely immoral.

In making genetic decisions to be effected by morally acceptable means, the benefits expected from a given course of action must be weighed against any risk (or loss of good) incurred. This is exactly the mode of moral reasoning used in deciding whether or not to use X rays in medical diagnosis, or radiation therapy in medical treatments. Should patients with cancerous growths, for whom (because of age and/or condition of health) the expectation of parenthood is quite small, be subjected to massive radiation therapy? The answer here is obviously affirmative. But how does one compare the detection of a case of tuberculosis by X-ray survey with the genetic harm that will befall someone in generations to come? How compelling should the indication be before an unborn child is subjected to damage by a fluoroscopic examination of its mother?[75] Moral reasoning that applies the principle of prudence, or the principle of proportion between effects both of which arise from a single action, is notoriously inexact. Still, it is certain that it is *immoral* to be imprudent, and it is a dereliction of duty not to make this sort of appraisal as best one can, and to act upon the best knowledge one can secure.

It is hardly utopian to hope that with the dissemination of genetic knowledge there will arise increased concern about this problem, and among an increasing number of people a far greater moral sensitivity to their responsibilities to the future generations of mankind. Such an "ethics of genetic duty" was well stated by H. J. Muller: "Although it is a human right for people to have their infirmities cared for by every means that society can muster, they do not have the right knowingly to pass on to posterity such a load of infirmities of genetic or partly genetic origin as to cause an increase in the burden already being carried by the population."[76]

There is ample and well-established ground in Christian ethics for enlarging upon the theme of man's genetic responsibility. Having children was never regarded as a selfish prerogative. Instead, Christian teachings have always held that procreation is an act by which men and women are to perform their duty to future generations of men. If a given couple cannot be the progenitors of healthy individuals—at least not unduly defective individuals—or, if they are the carriers of serious defect, then such a couple's "right to have children" becomes their duty not to do so, or to have fewer children. The science of genetics may be able to inform them with certain knowledge of the fact-situation. That would be sufficient to place eugenic reasons among those serious causes justifying the systematic practice of lifelong *un*parenthood, or of less parentage.

What is lacking is not the moral argument but a moral movement. The Christian churches have in the past been able to promote celibacy to the glory of God—men and women who for the supreme end of human existence "deny themselves" (if that is the term for it) both of the goods of marriage. These same Christian churches should be able to promote voluntary or "vocational" childlessness, or policies of restricted reproduction, for the sake of the children of generations to come. In place of Muller's "foster pregnancy," the churches could set before such couples alternatives that might be termed "foster parentage"—all the many ways in which human parental instincts may be fulfilled in couples who for mercy's sake have no children of their own. These persons would be called upon to "deny themselves" (if that is the term for it) one of the goods of marriage for the sake of that end itself. And they would honor the Creator of all human love and procreation, in that they would hold in incorruptible union the love that they have and the procreation they never have, or have no more.

Chapter 2: Shall We Clone a Man?

> This little pig built a spaceship,
> This little pig paid the bill;
> This little pig made isotopes,
> This little pig ate a pill;
> And this little pig did nothing at all,
> But he's just a little pig still.
> Frederick Winsor, *The Space Child's Mother Goose*

The remarkable advances in biochemistry and molecular biology in recent decades, and man's increasing skill in identifying hereditary diseases or defects, seem to hold out the prospect that in the future we may be able to inaugurate a program of positive eugenic improvement, by means of alterations to be made in our genes. This is called by various names, such as "genetic engineering," "genetic surgery," or "genetic alchemy."

Our modern scientific knowledge and the technical possibilities it affords seem also to have breathed new life into the idea that eugenic development and man's control of his evolutionary future may be based on some form of somatic selection. Earlier proposals for eugenic reconstruction and control based on somatic selection (selection of the *phenotypes* to be favored in human reproduction) could appeal only to tyrants, or to invincibly technocratic minds. These would have us adopt the procedures of animal husbandry.

It may be granted that in the case of some serious genetic defects *dominantly* transmitted the birth of individuals in the future having these defects might be prevented, or decreased in incidence, at comparatively small cost in the number of carriers and sufferers declared genetically dead. But in the case of serious genetic diseases *recessively* transmitted, no significant reduction of incidence in the population could be accomplished without decreeing the genetic death of vast numbers of people. The gene pool would have to be forced through a very narrow corridor of those allowed to reproduce, even to attain the goals of preventive eugenics. The more we learn about more diseases and their patterns of inheritance, the more radical the selection would need to be. Add to this the goal of reconstructing mankind by the selection of superior types to be favored in reproduction. Relentlessly, again and again, the gene pool would have to be forced through the same narrow corridor of those types chosen to be the sponsors and progenitors of the future generations of men. Apart from the difficulty of knowing who is the "good" or "qualified" man or what genetic strengths should be favored—and apart from the difficulty of knowing who is going to decide such questions—this would be an unconscionable procedure. In the language of ethics, even if somatic or phenotypical selection could be known to be a *good* thing to do (because of its consequences for future individuals and for the species), it would not be *right* (because of its massive assaults upon human freedom and its grave violation of the respect due to men and women now alive and to human parenthood as such).

In the face of this dilemma, genetic "alchemy" ("engineering" or "surgery") holds out the possibility that by altering our genes before they produce future genotypes, there may be a practice of genetic medicine (at least preventively) that will be good for our progeny and also could rightfully be performed upon living men and women.

Some geneticists, however, seem to regard this as a

remote prospect, or at least one that need not be waited for. They have, therefore, advocated—or suggested or pondered—schemes for the control and management of mankind's genetic future that can be described as a "mix" of some of the techniques made possible by present-day molecular biology, biochemistry, and genetics with the older notions of selecting phenotypes to be favored in human reproduction.

There are two such combination proposals. The late H. J. Muller was a passionate advocate of a program of positive genetic improvement through *voluntary* somatic selection, made possible today by eugenically motivated artificial insemination, by frozen semen banks, and by a continuing evaluation of and publicity about the donors available to recipients who wish to be our genetic pioneers instead of passing on their own precious genes. I have had occasion elsewhere to comment at length upon Muller's proposals (see chap. 1, above).

Clonal Reproduction as a Proposal for Man

I wish now to bring under serious scrutiny the suggestions and predictions made by Joshua Lederberg, Professor of Genetics and Biology at Stanford University, in some of his recent articles and columns. This is the proposal that *clonal* reproduction may be choice-worthy as a diversion to be introduced into human evolution, and a good way for man to control and direct the future of his species. Lederberg's notions combine somatic, psychosociological, and genetic criteria for the selection of the phenotypes to be perpetuated in reproduction with an arsenal of techniques—a more exquisite arsenal than the one from which H. J. Muller drew in choosing future possible applications of molecular biology. Lederberg

seems driven to his proposals because, like Muller, he
believes genetic surgery or alchemy to be a remote possibility, but also because of the fact that when genetic
reconstruction or repair is accomplished, it would then
take twenty years to determine whether or not the experiment had succeeded in producing a superior or even
an adequate phenotype. Therefore he, like Muller, believes we should begin with the existing types whose
strengths we know, and find a way to ensure their replication in greater numbers. Since the procedures he envisions using in the control of man's future seem as
remote as genetic engineering (or as imminent, in this
age of galloping scientific and technological progress),
Lederberg's crucial argument for clonal reproduction
must be that it would replicate already existing men
whose stature can now be measured and esteemed.

It is difficult to tell whether or not Lederberg advo-
cates the prospect which he has commented on in several recent articles.[1] He seems to disavow advocacy
when he concludes by saying that "these are not the
most congenial subjects for friendly conversation, especially if the conversants mistake comment for advocacy."[2] He seems to be pondering possibilities that
would be worthwhile if only there were some good reason for putting them into practice, and he expressly
claims simply that "it is an interesting exercise in social
science fiction [later: a "fantasy"] to contemplate the
changes in human affairs that might come about from
the generation of a few identical twins of existing per-
sonalities."[3] Still, on certain philosophies of the un-
avoidability of scientific and technological progress,
whatever can be done will undoubtedly be done. To
predict and even to ponder the prospects, while leaving
open the question of whether the experiment *should* be

tried, or having only relativistic ways of answering that question, amounts to much the same thing as espousal. A *determinism* in regard to the increasing application of medical technology combined with a radical *voluntarism* in regard to man's control of the future of his own species tends to erase the distinction between predictive comment and advocacy. Even if this is not a fair representation of Lederberg's position, there is ample espousal interspersed in his comments on the future possible uses of clonal reproduction.

"Clone" is a botanical term meaning "cutting"; the word "colony" is one of its cognates. Clonal reproduction means vegetative or asexual reproduction. Not all reproduction of vegetative forms is, of course, asexual or the result of cuttings or dispersal from a single source; but there are natural clones in plant life that seem to have advantages under certain luxuriant environmental conditions. Examples of clonal or asexual biological reproduction are the growth of an intact worm from each of the segments when an earthworm is cut in two, and the growth of identical twins from the segmentation of a single genotype in man.

Clones can be artificially created. This is the way chrysanthemums are produced commercially. In 1952 at the Institute for Cancer Research in Philadelphia Drs. Robert Briggs and Thomas J. King replaced the nuclei of freshly fertilized egg cells of the leopard frog *Rana pipiens* with nuclei from *blastula* cells (i.e. early embryonic tissue at the end of the cleavage stage of the development of a single individual of that species). They produced a clone of free swimming embryos (tadpoles) having the same genetic endowments as the tissue cell donor.[4] In 1956 embryonic clones were produced by using embryonic tissue older than the blastula stage.[5]

The objective of the experiments was to show the stability of the nuclear transplant and the hereditability of characeristics in cell development—for possible use of this knowledge in the study of cancer in various tissues. A few of the individual cloned tadpoles were allowed to grow to maturity in 1952 and 1956. Since *R. pipiens* takes three years from egg to egg, and thus would require both frog and experimenter to hibernate, interest in cloning adult individuals could proceed only by shifting to a species that has a shorter generation span and other characteristics that flourish under laboratory conditions. Thus in 1961 Dr. J. B. Gurdon, a zoologist at Oxford, produced a clutch of toads from the South African clawed toad *Xenopus laevis,* each toad having exactly the same genetic characteristics that the others and the cell-donor animal had—in fact a clone.

Such experiments have not yet been reported successful even for the laboratory mouse. Still, the possibility that "in a few years" the experiment can be performed upon man is the basis for the proposal that Lederberg contemplates applying in the artful management of human reproduction. At the moment, therefore, as we examine this possible diversion, or improving mechanism, to be introduced into human evolution, each of us can have in mind (*de gustibus . . .*): clones in vegetable life, the identical-twining of oneself, earthworms from cuttings, toads, or chrysanthemums. Dr. James F. Bonner of the California Institute of Technology believes that within fifteen years we will know how "to order up carbon copies of people"; he also speaks of this as "human mass production."[6]

Instead of "clonal" or "vegetative" reproduction, it would be more exact to call this: reproduction by enucleating and renucleating an egg cell that has already been launched into life by ordinary bi-sexual reproduction. The question, Shall we clone a man? means, Shall we renucleate the human fertilized egg cell?

Drs. Robert Briggs and Thomas J. King did this with frogs fifteen years ago, when they showed that it was possible by some deft microsurgery on *R. pipiens* to replace the nucleus of a fertilized egg with a nucleus taken from a tissue cell of another frog. The cell containing the genotype of the developing embryo—the cell which resulted from the normal chance combination of parental genes—was "renucleated," i.e. its nucleus was replaced by one having the genotype of a single already existing individual of the species. The new frog made by "nuclear transplantation" would have been the identical twin, a generation late, of but a single "parent." This was the fundamental experiment upon which Dr. Gurdon later built in actually producing a clutch or clone of toads.

The fact of genetics upon which the possibility of clonal human reproduction can be envisioned is, of course, the fact that our genotype exists in not only our egg or sperm cells, but also in every cell of every tissue of our bodies. There are two main technical difficulties that had to be overcome in experiments with toads, and would have to be overcome in experiments with other laboratory animals and then with man.

(1) Our genotypes (or genetic information) exist in different developmental "fixes" in the various tissues and organs of our bodies. We would not grow into whole, differentiated bodies unless the genes for eye color were somehow "switched off" in the nuclei of the cells of our finger nails, and the characteristics of the latter switched off in the nuclei going to cells in the eye. The switching off may be the work of special proteins

called histones. These must somehow be counteracted. The trick will be to discover how to decouple these switches, so that we can begin with a nucleus and genotype taken from specialized tissue cells, and enable them, when transplanted, to develop into a whole, organically differentiated member of the species. Alternatively, we might need to locate some special human tissue in which the cells have not become fully differentiated, or specialized, and therefore contain a greater degree of the totipotency of the original egg nucleus. Nuclei from early embryonic (blastula and early gastrula) cells in several species of amphibia have been shown to be totipotent. Dr. King reports that in *R. pipiens* it is now well established "that blastula and early gastrula nuclei, capable of promoting completely normal development, are undifferentiated and as such are equivalent to the zygote nucleus at the beginning of development." Following the gastrula stage, at least in *R. pipiens,* there appears to be a progressive restriction in the ability of cell nuclei to promote normal development. Dr. King writes that "the evidence at hand indicates that . . . chromosome abnormalities are not the result of technical damage sustained during the transfer procedures, but rather represent a genuine restriction in the capacity of somatic nuclei from advanced cell types to function normally following transfer into egg cytoplasm."[7] This restriction seems not to be the case, or not so extensive, in the toad species Xenopus, and Dr. Gurdon succeeded in producing adult toads when the new nucleus was taken from gut tissue at the late tadpole stage.

(2) The second formidable technical difficulty arises from interaction between the transplanted nucleus and the cytoplasm into which it is inserted. There may be a

number of reasons for nuclear-cytoplasmic incompatibility leading to abnormalities in development. If, for example, "the majority of donor nuclei were in a prolonged interphase condition, there would be an everdecreasing probability of finding in a random sample of cells nuclei that are ready to enter normally into the cleavage cycle of the egg."[8] The egg cell might cleave, causing disruption in the nucleus, and serious abnormality in development.

These are two of the problems that will have to be solved before man can give himself the choice of embarking upon clonal reproduction. Their solutions may require the combination of eggs renucleated from several different specialized human tissues in order to restore "the full range of developmental capability" which was in one's own original genotype, and in order to transmit this to another human being without availing oneself of the hoary old method of sexual reproduction. Yet Dr. Lederberg does not doubt that this will be accomplished, and in the time span of a few years.

Reasons Said to Favor Clonality

What good reasons are there for man's adoption of clonal or vegetative reproduction when it becomes feasible? Why should we clone a man—or colonies of such men? This is the chief issue.

Lederberg offers a number of reasons, both individual and social, why clonal reproduction might not be a bad idea.

In the first place, "unlike other techniques of biological engineering there might be little delay between demonstration and use. . . . If a superior individual—and presumably, then, genotype—is identified, why not copy it directly, rather than suffer all the risks . . . involved in the disruptions of recombination [?]" In this rhetorical question, Lederberg prosecutes his case for the comparative advantage of clonal reproduction over "genetic engineering." In so doing, he stresses the hazards of making mistakes in reconstructing genes, and forgets for the moment the hazards (which he himself later stresses) in cloning a man or large parts of mankind. "What to do with the mishaps," he later writes, "needs to be answered before we can undertake these risks in the fabrication of humans"—i.e. before implanting "a special nucleotide sequence into a chromosome." Our genetic system is so complex that experiments in the surgical repair of the system are "bound to fail a large part of the time, and possibly with disastrous consequences if we slip even a single nucleotide."

Yet it is evident that Lederberg is not deterred from experimentation by mishaps, or by the problem of what to do with them. What he refuses to endorse is simply waiting twenty years "to prove that the developmental perturbation was the intended, or in any way a desirable one." The "inherent complexity of the system" does not exclude experimentation, and many a mishap, on the way to clonal reproduction—the developmental perturbation and diversion to be introduced in the control and redirection of the future of the species. What is excluded is simply "any *merely prospective* experiment in algeny [genetic alchemy]" (italics added).

In other words Lederberg's premise at this point is simply expounded in the question (which, if an argument, is circular): "If a superior individual—and presumably, then, genotype—is identified, why not copy it directly, rather than suffer all the risks . . . involved in the disruptions of recombination [?]" I say circular, because the risks may be sufficient or of *such a kind* as to rebut both sorts of experimentation upon human generation.

In the second place, there are individual decisions and grounds for these decisions which, Lederberg believes, will be sufficient to insure clonality, at least as a minority practice once it is feasible, and "given only community acquiescence or indifference to its practice." For example, if one is the carrier of a serious recessive disease, he could twin from himself another carrier of the same serious recessive disease, rather than risk having an overtly defective offspring by mixing his genes with those of his wife (who might also be a carrier). He might either make sure of an exact copy of himself or he could copy his spouse; and in either case could attain "some degree of biological parenthood." "Copying one's spouse" would also be a way of predetermining that the offspring will be a boy or a girl (whichever is desired); "nuclear transplantation is one method now verified to assure sex control, and this might be sufficient motive to assure its trial."

At this point Lederberg cannot avoid introducing a possibly dysgenic consequence of the adoption of clonal reproduction for the foregoing reasons even as a minority practice—although his opening words disavow disapproval. "Most of us pretend to abhor the narcissistic motives that would impel a clonist," he writes, but then goes on to state the genetic truth: "He—or she—will pass just that predisposing genotype intact to the clone." From this the immediate inference is: "Wherever and for whatever motives close endogamy has prevailed before, clonism and clonishness will prevail." Clonal reproduction will make for clanishness. We shall return to the question this raises—whether clonal reproduction would not in fact be a genetic disaster, because it would be a step backward from "adaptability," which is "man's unique adaptation."

Thirdly, clonal reproduction might have considerable medical and social advantages. It would be a backup procedure for building men with interchangeable parts. The members of a colony would all be identical twins, and this would make possible among them "the free exchange of organ transplants with no concern for graft rejection." More important would be the societal benefits. "Intimate communication" among men would be notably increased. Misunderstanding would be eased, and there would be better social cohesion and cooperativeness among us. This expectation is based upon the observation that "monozygotic twins are notoriously sympathetic, easily able to interpret one another's minimal gestures and brief words." We could produce pairs or teams of people for specialized work in which togetherness is a premium: astronauts, deep-sea divers, surgical groups. Clonal reproduction would afford us a solution to the "generational gap"; "it would be relatively more important in the discourse between generations, where an elder would teach his infant copy."

Lederberg grants that he knows of no "objective studies" of the "economy of communication" among monozygotic twins. Nor does the present writer. It would seem, however, that Lederberg has not had the experience of being the father of either monozygotic or heterozygotic twins. The crude evidence in human experience does not lend unequivocal support to the expectation that "intimate communication" would be increased and not also ambivalence and animosity in personal relations. Growing up as a twin is difficult enough anyway; one's struggle for selfhood and identity must be against the very human being for whom no doubt there is also the greatest sympathy. Who then would want to be the son or daughter of his twin? To mix the parental and the twin relation might well be psychologically disastrous for the young. Or to look at it from the point of view of parents, it is an awful enough responsibility to be the parent of a son or daughter as things now are. Our children begin with a unique genetic independence of us, analogous to the personal independence that sooner or later will have to be granted them or wrested from us. For us to choose to replicate ourselves in them, to make ourselves the foreknowers and creators of every one of their specific genetic predispositions, might well prove to be a psychologically and personally unendurable undertaking. For an elder to teach his "infant copy" is a repellent idea not because of the strangeness of it, but because we are altogether too familiar with the problems this would exponentially make more difficult.

In the fourth place, finally and for completeness' sake, it should be added that scientific interest is in itself for Lederberg a sufficient motivation for cloning a man, given community acquiesence or indifference to the project: "we would at least enjoy being able to observe the experiment of discovering whether a second Einstein would outdo the first one."[9] But of course it would take more than twenty years to determine that.

Reservations and Objections

Lederberg himself brings up the most serious genetic objection to switching to clonal reproduction. "Clonality as a way of life in the plant world," he writes, "is well understood as an evolutionary cul-de-sac. . . . It can be an unexcelled means of multiplying a rigidly well-adapted genotype to fill a stationary niche. So long as the environment remains static, the members of a clone might congratulate themselves that they had indeed won a short-term advantage." But it is, Lederberg says, "at least debatable" whether "sufficient latent variability," now insured by human sexual reproduction, would be preserved "if the population were distributed among some millions of clones."[10]

This therefore is not Lederberg's overall eugenic proposal; it is rather "tempered clonality" that he thinks might be desirable and workable. "Tempered clonality" would be an attempt to have the best of both worlds: asexual reproduction for uniformity, for intimate communication, and for multiplying proven excellence; sexual reproduction for heterogeneity and innovation. "A mix of sexual and clonal reproduction makes good sense for genetic design. Leave sexual reproduction for experimental purposes; when a suitable type is ascertained take care to maintain it by clonal propagation."

Horticultural practice, Lederberg writes, "verifies" that this "mix" or "tempered clonality" would be the best genetic design. Is it not also evident that to maintain the mix, to *temper* clonality so that the cul-de-sac of genetic rigidity can be avoided, will require careful and compulsory management of the two groups of people (or the two groups of reproductive events)? Some must be chosen to clone; others to engage in sexual reproduction. It may be that clonality "answers the technical specifications of the eugenicists in a way that Mendelian breeding does not" (because the latter continues to mix good with bad genes in sexual reproduction, while the former proceeds at once to replicate desired genotypes directly again and again). Still this mix of the best of two worlds picks up the worst of one: dictated breeding or nonbreeding patterns. This grand genetic design for avoiding the dysgenic consequences of switching altogether to clonal replication necessarily involves managers decreeing the large numbers who shall reproduce sexually and those who shall clone. The horticultural practice that verifies that a mix of the two makes the best genetic design also verifies that this can be instituted and maintained only by the power to decree: "Sexual Reproduction Forbidden." If this design does not exceed human wisdom, it certainly falls below the morally permissible.

It might also be asked what happens when one fine day a member of one of the clones decides (as a Dostoevski might imagine) to stick out his tongue (or a more appropriate organ) at the whole scheme? What if he and a number of his fellow clonites find partners (in a neighboring clone having the opposite sex, or in the ordinary population) willing to join them in setting a new fashion in reproduction (this time, the hoary old method you can read about in medical text books)? Would not all the recessive deleterious genes, plus the mutations that have been going on in clones for years, be suddenly dumped upon the world's population?

Perhaps this objection need not be raised; but if not, there would be another crucial objection—depending upon whether the human clonites we succeed in producing in the future are sterile or fertile in (forbidden) sexual reproduction. Although many of Dr. Gurdon's cloned toads were fertile, a higher proportion were sterile than among frogs of this species *reared in the laboratory* and a still higher proportion than in a normal population of "wild type" *X. laevis.*[11] So it may be that the eugenic managers of the "mix" (*tempered* clonality) need not raise the flag "Sexual Reproduction Forbidden" for all, but only for *some* (possibly a good number) of the clonites. Perhaps, instead, a somewhat larger proportion of the population of clonites than is the case in the ordinary "wild type" human population will be sterile. In the case of these, there need be no fear of dysgenic consequences if clones begin once again to practice sexual reproduction. But then we have come upon a point which Dr. Lederberg neglects to mention: A step upon this track may be irreversible from the first clone onward, for a large minority. Once having adopted vegetative reproduction, it may be *that* or no reproduction at all, for a significant proportion of the preferred types.

Dr. Lederberg concedes that, "even when nuclear transplantation has succeeded in the mouse, there would remain formidable restraints on the way to human application, and one might even doubt the further investment of experimental effort." He has in mind technical difficulties, which to the layman seem about as formidable as those confronting future possible applications of genetic surgery. Also and at the same time, he comes to the question on which he says he and his colleagues differ widely, namely, whether "anyone could conscientiously risk the crucial experiment, the first attempt to clone a man." Again to the layman, this moral question seems as formidable as the "disastrous consequences if we slip even a single nucleotide" to which Lederberg appealed in opposing genetic engineering as the first choice procedure.

Dr. Thomas J. King was careful to explain that in the experiments performed by himself and his colleague Dr. Robert Briggs, in which frog egg cells were successfully

renucleated, this was done from *cells taken from embryonic tissue.* The nucleus was not taken from (in
every sense of the word) *embryonic* cells. The nuclei
came from a population of cells, and cells in a population or in a tissue may not have the same degree of
development. No doubt these were at least *morphologi-
cally* differentiated tissue cells (differentiated in appearance, having a ciliated brush border), but the cells used
may not yet have been fully *chemically* differentiated
embryonic cells. Thus they may have been more potent
cells, not yet fully specialized, with only some of the
potentialities that were in the original genotype
switched on at that point on the way to becoming embryonic tissue cells.[12] However, Dr. J. B. Gurdon's success in growing individual frogs from renucleated egg
cells seems clearly to have been from *differentiated tissue cells,* not from indeterminate cells in a population.[13] It is notable, however, that he began with tissue
cells from the intestines in the tadpole stage. Not just
any tissue cells, or at any stage of the individual's development, would have worked. These considerations are
sufficient to indicate the technical difficulties in the
way of successfully cloning laboratory mice, other animals, and finally man—which make even Lederberg
grant that "one may doubt the further investment of
experimental effort."

The formidable moral question concerning "the crucial experiment, the first attempt to clone a man" can
also be made clear from these earlier experiments. Much
depends upon whether the experimenter happens to get
the nucleus from a tissue cell (whether intestinal or one
of a population that is still totipotent) at just the right
moment *in its development* when he places it in the
fertilized egg cell. As I understand it, if the nucleus from

the donor tissue cell is in a prolonged interphase condition while the egg cell into which it is inserted is undergoing cleavage (nuclear-cytoplasmic incompatibility; see
n. 8 and pp. 67-68, above), the result could be a *monstrosity,* and not normal embryonic development (King-
Briggs) or an individual with the full characteristics of
others of the species (Gurdon). That would seem to
confront the reproduction of a man by clonal replication with essentially the same *sort* of hazard, and the
same formidable moral objection, as confront the slipping of a single nucleotide in the repair or reconstruction of genes.

Lederberg supposes that the first crucial attempt to
clone a man may be postponed "until gestation can be
monitored closely to be sure the fetus meets expectations"—or until "extracorporeal gestation," permitting
the same surveillance, is perfected in laboratories. This
obviously connects the question involved in cloning a
man—what to do with resultant mishaps—with the question of whether we are going to make objects of scientific experimentation out of nascent human lives *in
utero* or in laboratory conditions simulating this condition, and of the natural or artificially induced mishaps
of such undertakings. In any case, the moral question
turns largely upon how we are going to "make sure the
fetus meets expectations," and by what standards of
judgment, warrants, or ethical points of reference.

Before asking whether Lederberg has any very good
answer to that question, it should be pointed out that,
while waiting to see whether anybody is going to venture to clone himself or another congener, scientists will
reduce the difference between trials with animal egg
cells in laboratory cultures and the future possible event
of cloning a man. "Human nuclei, or individual chromo-

somes and genes, will be recombined with those of other animals species; these experiments are now well under way in cell culture. Before long we are bound to hear of tests of the effect of dosage of the human twenty-first chromosome on the development of the brain of the mouse or the gorilla." Somewhat triumphantly Lederberg adds: "The mingling of individual human chromosomes with those of other mammals assures a gradualistic enlargement of the field and lowers the threshold of optimism or arrogance, particularly if cloning in other mammals gives incompletely predictable results."

This brings us again to the same formidable moral objection. In the case of cloning a man, the question is what to do with mishaps, whether discovered in the course of extracorporeal gestation in the laboratory or by monitored uterine gestation. In case a monstrosity—a subhuman or parahuman individual—results, shall the experiment simply be stopped and this artfully created human life killed? In mingling individual human chromosomes with those of the "higher" mammals (given sufficient dosage and "a few years"), what shall be done if the resulting individual lives seem remarkably human? Moreover, Lederberg does not contemplate only experiments "augmenting" animal cell cultures with "fragments of the human chromosome set." The reverse is also to be done. "Clonal reproduction, and introduction of *genetic material from other spheres*" are two paths already opened up in *human* evolution (italics added). Lederberg "infers" these twin genetic policies instead of taking the other, somewhat longer road of genetic engineering. But surely these are paths no less fraught with mishaps knowingly if not intentionally created. We must face the grave moral question of what to do with them.

On this fundamental question it must simply be said that Lederberg is not very helpful. He calls for reasoned and sensitive moral judgment on these matters. He fears that the issue will be settled adversely if the public gets wind of what is now or (he believes) will shortly be going on in laboratories, or because of the publicity concerning the first results of cloning a man or of giving the human chromosome set large dosages from "other spheres." "The precedents affecting the long-term rationale of social policy will be set," he concludes, "not on the basis of well-debated principles, but on the accident of the first advertised examples. The accidents might be as capricious as the nationality, the batting average, or public esteem of a clonant; the handsomeness of a para-human progeny; the private morality of the experimenters; or public awareness that man is a part of the continuum of life."

The accident of the first examples (advertised or not) and the probability of a great many of these accidents are the questions now under examination. The "accidentality" of how the question of these accidents might be settled by an unthinking public is not at issue. We may grant the nonrationality and the capriciousness of settling any human moral public policy question by any of the foregoing criteria (including the last one: the fact that man is a part of the continuum of life). However, it is just as nonrational to refer, as Lederberg does, to "the *touchiness* of experimentation on obviously human material" (italics added). The moral issue to be raised concerning "sub-human" hybrids and concerning mishaps in applications of molecular biology is precisely whether this "touchiness" is not entirely justified. Indeed, is this not a sign of a moral prohibition?

Lederberg begins by announcing quite correctly that "scientists are by no means the best qualified architects of social policy,"[14] and by limiting their function to the essential tasks of interpreting the technical challenges facing humanity and having the forethought to determine various scientific efforts that might be put forth to meet these challenges.

But when he comes to the questions of purpose and value that must provide the foundation of any public policy, it is still a purely scientific value that heads all the rest, and a test most likely to appeal to "sceptical scientists." It is not anything morally substantive about man's nature, himself, or human purpose, but "the growing richness of man's *inquiry about* nature, about himself, and his purpose" (italics added). That indeed is the arch-scientific value. Despite his previous announcement, Lederberg makes this the overriding value. Interspersed is some truly admirable moral rhetoric about how scientific inquiry leads to "a humble appreciation of the value of different approaches to life and its questions, of respect for the dignity of human life, and of individuality." This is plainly the "humility" that will keep open the options and answer no moral question in the area of eugenics, until the scientific accounting of our genetic options is complete. It is also one that will never find any moral limit upon the choice among options. Instead there will be only limits derived from a "rigorously mechanistic formulation" of the life sciences. This view would seem to be the reason for Lederberg's decrying "the arrogance that insists on an irrevocable answer to any of these questions of value."

On the basis of this article alone, it is evident that Lederberg has made his choice between humane civilization and a civilization based on scientific-mechanistic values alone (if the latter is a proper manner of speaking). "Humanistic culture," he writes, "rests on a definition of man which we already know to be biologically vulnerable." This is a choice made not without trouble by Lederberg the man, since he knows that "the goals of our culture rest on a credo of the sanctity of the human individual." Despite this, however, Lederberg the scientist can only pose the question that for him has but one obvious answer: "How is it possible for man to demarcate himself from his isolated or scrambled tissues and organs on one side, and from experimental karyotypic hybrids on the other?"

The entire proposal that we should clone a man and proceed with mixing chromosomes from "other spheres" with human material is, therefore, simply an extrapolation of what we should do from what we can do. There are really only technical questions to be decided. At root, therefore, the proposal denies that the mishaps constitute a crucial moral problem. The mishaps are only misadventures that may or may not awaken in a capricious public a latent humanism sufficiently powerful to stop the experiment.

Specifically concerning the mishaps, Lederberg says only that *"pragmatically,* the *legal* privileges of humanity will remain with objects that *look* enough like men to grip their *consciences,* and *whose nurture does not cost too much"* (italics added). (The reader is told later that "the handsomeness of a parahuman progeny" would be a capricious test, as indeed it is.) At this point Lederberg's own latent humanism rises to the surface of his scientific pondering of the options, and he is impelled to write: "Rather than superficial appearance of face or chromosomes, a more rational criterion of human identity might be the potential for communication within the species, which is the foundation on which the unique glory of man is built." With this statement the paragraph once ended; and it seems a minimum statement of a basis for the respect and protection of human life. But this statement was promptly corrected by a footnote in the version of this article which appeared in *The American Naturalist,*[15] and by the addition of a sentence in the text of the article for the *Bulletin of the Atomic Scientists.* The footnote stated: "On further reflection I would attack any insistence on this suggestion (which I have made before) as another example of the intellectual arrogance that I decry a few sentences above—a human foible by no means egregious." This judgment was made a part of the text in the later revision: "Insistence on this suggestion, of course, would be an example of the intellectual arrogance decried above." Thus Lederberg corrected his announcement of a possibly human criterion for demarking man from other genetic spheres and from artificially scrambled genes.

In the case of the mishaps, therefore, we are left with only the pragmatic problem of agreeing upon the line to be drawn empirically between those that *look human* enough and those that *look* subhuman or parahuman. Given this, we could proceed to manufacture, select, and eliminate progeny. This is simply a way of saying that the mishaps do not constitute a moral problem, and that the management and control of human evolution and of the future generations of men have no intrinsic limits. The procedure has only consequences, and goals "read in" to the object-matter by the experimenter and our future possible public policy managers.

The same muddled moral reasoning is to be found in the comments Lederberg made upon the problem of

abortion in a colloquium at a session of the American
College of Physicians in 1967. He rightly observes that
"unless we learn to deal with this contemporary issue in
a humane way, we will never in the future be able to
cope with the subtler problems of qualitative intervention in the finer points of human reproduction."[16] The
same perspectives we have examined above are applied
to this currently pressing issue; and the yield is no greater, for anyone moderately accustomed to the logic and
problems of ethics.

Lederberg states that the question "When does life
begin?" can have no answer "apart from the purposes
for which the question is raised." This is a fine, formal
requirement of moral discourse; and of course the purpose for which the question is raised in the matter of
abortion is to determine when human life has rights in
exercise and is a life due to be given respect and protection as a congener. But Lederberg proceeds to ask and
answer the question with quite another purpose in
mind—which is all right, but not (according to Lederberg's own formal requirement) in the present context.
"Life is a continuum: If life had a beginning at all, it
was an event that occurred some 3 billion years ago."
Thus Lederberg gets back the answer *he* prejudged, by
forgetting that the question has different meanings in
different contexts. This leads to irrelevancies (in the
context of the purpose for which the question of abortion is raised) such as: "No more than a small percent of
the total nucleotide composition of the diploid human
nucleus differentiates the human being from the ape, or
the monkey, or other primate species." Argument from
the continuum of life (which was not in question) is
sufficient to place the genetic material which continued
over the course of 3 million years, human sperm, ovum,

fetal life, and at least the newborn human being all on
the same level. (This is remarkably like the identification of the problem of abortion with that of contraception in some conservative Roman Catholic circles—to
the conclusion that sperm and ovum, the zygote and
embryo, should alike be protected from artificial assault.)

The analogy between Lederberg's remarks on abortion
and his views on the "finer points" of intervention in
human reproduction becomes most clear when he remarks—not incorrectly if communication is the test—
that:

> An operationally useful point of divergence of the
> developing organism [in the whole continuum] would
> be at approximately the first year of life, when the
> human infant continues his intellectual development,
> proceeds to the acquisition of language, and then par-
> ticipates in a meaningful cognitive interaction with his
> mother and with the rest of society. At this point only
> does he enter into the cultural tradition that has been
> the special attribute of man by which he is set apart
> from the rest of the species.[17]

But again Lederberg draws back from having drawn a
human and a moral line: this time in regard to the treatment to be accorded the newborn, above in regard to
what we should do with clonal mishaps. "I do not advocate a discussion of infanticide," he says, having just
begun a discussion at its most essential point (if the
point is ethically correct). Just as he drew back to those
artifacts of the proposed applications of molecular biology that/who *look* enough like men to grip our consciences (emotions?), so here he observes that "we are
all so emotionally involved with infants that this in itself

is enough to create an inevitable and a pragmatically useful dividing line [at birth]."

Lederberg has therefore provided himself with no intellectual foundation for immediately making the following *dictum:* "To discuss the fetus during prenatal life as if he were a human being is merely to reflect the emotional involvement of that observer." Surely he had just appealed to the same sort of emotional involvement with another life during that part of the continuum from birth to age one—as the only ground for not practicing infanticide. Of course, the latter emotional involvement is that of most of mankind, while the former may be "a set of tastes not shared by the majority."

Then is Lederberg recommending that moral questions and public policy be settled by polls of people's emotional involvement with prenatal life, with babies in incubators, with babies after normal birth, with experimental karyotypic hybrids?

No. On a generous interpretation, none of these "sets of tastes" (and not simply the one he opposes) should "be confused with an objective biological standard by which we can set up principles of social order as criteria for the operation of law." Since, however, from this no objective ethical standards have been shown to be forthcoming—and since to insist that the test should be a potentiality for communication within the species would be an instance of "arrogance" insufficiently informed of the continuum of life—Lederberg is driven in the end to hortatory appeals: "Help put this [abortion] issue on the plane where it deserves to be discussed: What are the realistic consequences of revision of the law for human welfare in the light of your own knowledge of the continuity of biological life and evolution?" In other words, while Lederberg does not want only scientists to make public policy, he wants only a scientific judgment to be made.

In the discussion following the colloquium, Lederberg was asked whether he would approve of infanticide in the case of monstrous births, or of allowing them to die.[18] His answer combined a genuine concern for human welfare ("Before I would advocate killing even monstrous births, I would want to inquire what the effect might be on the standards of care for other infants.") with opting for the time of birth as the significant dividing line (we should "do as much as we can to bring about the earliest possible detection of aberrations so that these genetic deaths can be made to occur at the period where they would have the least strenuous consequences for the other members of our society"). As for the aberrations that would still sometimes be born, his inclination would be that instead of killing them or letting them die, "they may be such interesting objects for humane observation and experimentation that it may be very well worth making very great efforts to keep them alive once they have started to exist." One may ask why, on Lederberg's principle of continuity, all lives before "they have started to exist" (i.e. been born) might not be fit subjects of medical experimentation?

The genetic possibilities I have been discussing mark an extraordinary "borderline" to which human ingenuity has brought mankind in the present day. Before stepping across that borderline we should let it teach us something. The borderline, as Helmuth Thielicke has written, is "the truly propitious place for acquiring knowledge"[19]—ethical knowledge. The genetic proposal to clone a man, and the minority practice of artificial insemination from a nonhusband donor, are borderlines that throw into bold relief *the nature of human parent-hood* which both place under assault. If cloning men had no other consequences, this alone would be suffi-cient to fault it. Our consideration of the proposal should enliven in our minds an important feature of parenthood which serves "to make and to keep human life human."[20] Any fundamental assault upon this is an assault upon the human and the personal element in parenthood.

Man is an embodied person in such a way that he *is* in important respects his body. He is the body of his soul no less than he is the soul (mind, will) of his body. There are more ways to violate a human being, or to engage in self-violation, than to coerce man's free will or his rational consent. An individual's body, including his sexual nature, belongs to him, to his *humanum*, his per-sonhood and self-identity, in such a way that the bodily life cannot be reduced to the class of the animals over which Adam was given unlimited dominion. To suppose so is bound to prove antihuman—sooner than later.

The nature of human parenthood may be summed up by saying that conjugal intercourse is a life-giving act of love-making or a love-making act of life-giving. Procrea-tion and the communications of bodily love, nurturing and strengthening the bonds of life with life, belong indefeasibly together—not, it is true, in every act of mar-riage—but between the two persons who are married. From their one-flesh unity together comes the one flesh of the child. Not, of course, because of any materiality genetics could prove or disprove, but because the com-munications of love and parenthood are both deeply personal acts and relations. This is all the more so be-cause they are not done exactly rationally. More of the selves are engaged because this bodily presence and these bodily events are not directed by the dominating rational will alone. Whatever substrata of uses there may be in sexual energy beyond the needs of the species, and whatever the mechanics of heredity, these are subsumed onto a human plane in conjugal love and in parenthood.

We procreate new beings like ourselves in the midst of our love for one another, and in this there is a trace of the original mystery by which God created the world because of His love. God created nothing apart from His love, and without the divine love was not anything made that was made (John 1). Neither should there be among men and women, whose man-womanhood (and not their minds or wills only) is in the image of God, any love-making set out of the context of responsibility for *pro-creation* or any begetting apart from the sphere of human love and responsiveness. Thus is our man-woman-hood created in covenant and for covenant—the covenant of marriage and the covenant of parenthood.

Men may be able to subdue the mystery of procrea-tion, they may be able to subdue all the wonders of human sexual response, in their sciences. But they can-not subdue the mystery in the fact that eminently human communications of marital love are also the places where we engage as pro-creators, and establish and step into the covenant of parenthood. Men can only deny that there is any mystery to be honored here; they can only reduce the matter to an accident of biological na-ture that could as well not have been so, or could be changed to vegetative reproduction. Herein men usurp dominion over the human—the dominion they hold rightfully only over the animals. This is bound to pierce the heart of the *humanum* in sex, marriage, and genera-tion.

The "experiment" involved in the thought, Shall we

clone a man?, and the technical possibility of doing so on a vast scale, may provide the truly propitious opportunity for acquiring the knowledge that the link between sexual love and procreation is not in us a matter of specific or animal consequence only, but is of truly human and personal import. To put radically asunder what nature and nature's God joined together in parenthood when he made love procreative, to disregard the foundation of the covenant of marriage and the covenant of parenthood in the reality that makes for a least minimally loving procreation, to attempt to soar so high above an eminently human parenthood, is inevitably to fall far below—into a vast technological alienation of man. Limitless dominion over procreation means the boundless servility of man-womanhood. The conquest of evolution by setting sexual love and precreation radically asunder entails depersonalization in the extreme. The entire rationalization of procreation—its replacement by replication—can only mean the abolition of man's embodied personhood.

Another thing we can learn from clonality as a thought-experiment, or as a borderline prospect, is a proper historical perspective upon the stress on procreation as the chief end of marriage in ages past. Never before in our history has breeding been so highly exalted as the chief end—indeed the isolated end—to be pursued by marriage as an institution, or by surrogate institutions of our own devising. Never was breeding so far separated from personal, human love. Never was breeding (or replication) set so far afield from the relations in which men and women express *fides* toward one another (not even in times past when *fides* meant only something negative or remedial). This is done from one side by those of our contemporaries who would justify

acts of sexual love beyond the sphere of responsible
procreation (by definition, marriage)—which then is
more and more for the procreation and education of
children *only*. It is done from the other side by any
scheme for the entire rationalization and control of "reproduction" (itself a metaphor of a machine civilization,
and one that already gets rid of the persons, and may,
by a next step, suggest getting rid of one of the "pro-
ducers" by means of clonal replication).

Up to this point we have, so to speak, been doing
genetic ethics. We have been examining a specific eugenic proposal in the light of some of the moral norms
that generally commend themselves to rational moral
agents—norms in which we may be instructed by the
borderline case of clonal reproduction when it is seriously put forward as a project worthy of man.

We need, finally, to stand at a distance from these
detailed incriminations or commendations of clonal reproduction, in order to examine its chief philosophical
underpinning. There is a total life-view at work in the
grand design of mixing clonal with sexual reproduction
in the control of man's own evolution. This same outlook is at work also in projects for positive or progressive eugenics and the redirection of the human future by
means of the "chemical control of genotypes"—"genetic
engineering."

In fact it may be said that the ethical violations we
have noted on the *horizontal* plane (coercive breeding or
nonbreeding, injustice done to individuals or to mishaps,
the violation of the nature of human parenthood) are a
function of a more fundamental happening in the *verti-
cal* dimension, namely, hubris and "playing God."

These, of course, are not very helpful characterizations of the attitude that is in control of some, if not all, of the progressive eugenic proposals that have been put forward in the modern age. We shall have to see if the fundamental outlook I have in mind can be analyzed somewhat more closely, and ask what then we are to think about this outlook. In undertaking to do this, I shall again make use of Lederberg's article. It would not be fair, however, to quote him on this point if it were thought that these viewpoints are in any way especially his. They are rather simply symptomatic of modern thought-forms in general (including some recent religious speculations that have placed all religious impulse behind the triumphalism of secular man, not to his essential correction or to critical judgment upon our inveterate utopianism). The following citations from Lederberg are, therefore, simply prisms through which can be clearly seen the operating, unspoken premises of modern man in general. These must be called radically into question.

The first of these premises is the replacement of categories appropriate to the ethical evaluation of behavior or of moral agency by categories appropriate to the elaboration and evaluation of *designs*. This is why Lederberg can so readily speak of "replacing evolution by *art*" (italics added), and proceed on the assumption that the means that are in the offing in "the geneticist's *repertoire*" (italics added) shall come into use. This is because only ends external to designs control them, and only the self-elected purposes of engineering place any limit upon what we shall undertake to do to pliable matter, restricted only by what that matter yields to our purposes.

To the contrary, in all the ages of man founded at all upon moral premises it has been silently if not explicitly assumed that "making" is quite different from "doing." The exercise of reason in practical applications of "art," in making designs, or in engineering consequences is one sort of human activity. The exercise of practical reason in political and moral agency ("doing") is another. The homeland of ethics is in the latter, not in the former or in any of its analogies in the conduct of human affairs, the "management" of crises or of man's own future. For man the doer, not for man the maker, genuine moral relations and moral considerations come into view. Instead of saying that modern man is prone to exercise an unlimited hubris or "play God" over the future, it is better to say that he thinks of himself as simply the engineer of the future—perhaps a humble maker artfully designing himself and his kind. From this, ethics has already been excluded, and not only by the Dionysian types.

The second premise is a more paradoxical one, and at the same time more widely and profoundly symptomatic of the modern condition. This is the combination of *boundless determinism* with *boundless freedom* in all our thoughts. Mechanism and arbitrary freedom go together. This is why Lederberg can correctly write that "in the last decade, molecular biology has given us a mechanistic understanding of heredity"—with the unproved assumption that this is all there is to be said about the destiny that underlies human freedom and calls it into play. And he continues by saying that this same molecular biology has given us "an outline of the methodology needed to do the same for development"—as if that is all that need be said about the absolute, unrestricted freedom with which man can now step forth to impose control over the future by making use of the outline provided by our mechanistic understanding of heredity.

There is no end, no limit to the mechanism, only an immanent determinism. There is no end, no limit to human freedom, only choices immanent to freedom's own election (plus the recalcitrance, if any, to be found in the outline of future possibilities). Looking into an open future, Lederberg can write that we now have "the new evolutionary theory needed to fashion a self-modifying system that plans, however imperfectly, its own nature." And we have seen how extensive is the contemplated self-modification. As there is no destiny overruling man's rulings, so there are no essential limits from humanistic values not already vulnerable to biological determinism. Strangely, heredity mechanistically understood has fashioned a limitless self-fashioning system (man), however imperfect the self-making may turn out to be.

The boundless determinism and the boundless freedom contained in this thought are not solely or mainly a product of science. They are rather a widespread cultural phenomenon or thought-form characteristic of man in the modern period. Dostoevski discerned this to be true. Where there is no God, no destiny toward which men move and which moves in them, then self-modifying freedom must be the man-God. This is necessarily correlated with the unlimited subjugation of everything else—of human reality as it now is, and of the genetic future—to the determinations of that usurped and absolute freedom.

It does not matter whether this is said with Dionysian exuberance or with the studied indifference of a scientific mind. The mind in any case floats freely over everything that is and contemplates that it may not be, or may be altogether otherwise. It thinks to create something out of nothing, or out of our chance and random chromosomes. Thus, in regard to possible reactions to the "fantasy" of switching to clonal reproduction, Lederberg observes that "if sexual reproduction were less familiar we might make the same comment about that."[21] He is "more puzzled by the rigor with which asexual reproduction has been excluded from the vertebrate as compared with the plant world" than he will be by man's achievement in changing all that. What can be the explanation of this puzzling penchant for sexual reproduction? "That clonal reproduction is mainly confined to plants," Lederberg writes, "may be a mere accident of cell biology." But whether it is accident or mechanistic determination that created us this way, there is in any case no reason for man to respect and honor any feature of the creation. He is now absolutely free to change all that—to choose to become his own self-creator.

Lederberg briefly tells his own myth for this. "In mythical terms, human nature began with the eating of the fruit of the Tree of Knowledge." In the original myth that meant the knowledge of good and evil; for Lederberg it means the knowledge of the accidents of cell biology or of the determinism disclosed to us by molecular biology and the laws of heredity. This is the way we have come to understand the garden of life in which man's freedom is planted. The functionally correlated statement must immediately follow: "The expulsion from Eden only postponed our access to the Tree of Life." For primitive man, that meant a natural immortality. For Lederberg it means neither immortal nor creaturely life, but *dominion* of man over man self-determining the conditions of future human life, the terms upon which future generations of men and our species in general shall live and move and have their being. When future generations say to us, Where were you when the foundations of our *humanum* were laid? the answer to be given them is quite evident—provided one believes that human freedom should now lay hold of godhood.

We may make the claim that the philosophy (not the science) underlying all this is untrue. This may take some temerity in the present age, but it is a not unreasonable assertion. We can say that, as for me and my house, we will continue to practice genetic ethics and genetic counselling and the elimination of grave defects in the simple service of creaturely life. This, I say, may require some temerity, but it is not unreasonable. In any case, there is no reason for anyone to be frightened out of his wits by the relativists, subjectivists, and sceptics who in moral science have not been able to make their case. Nor should anyone be frightened out of his ethical wits by grand eugenic designs.

It may take some temerity to oppose these grand interferences for man's self-reconstruction and control over the evolutionary future, but this is a not unreasonable position. In the present age the attempt will be made to deprive us of our wits by comparing objections to schemes of progressive genetic engineering or cloning men to earlier opposition to innoculations, blood transfusions, or the control of malaria. These things are by no means to be compared: the practice of medicine in the service of life is one thing; man's unlimited self-modification of the genetic conditions of life would be quite another matter.

Nor does it suffice to say that we are already introducing vast changes in the environment, and that these environmental changes are in fact altering mankind's genetic basis. Even if it is true that "the jet airplane has already had an incalculably greater effect on human population genetics than any conceivable program of calculated eugenics would have," that is an argument which cuts both ways. Mankind has not evidenced much wisdom in the control and redirection of his environment. It would seem unreasonable to believe that by adding to his environmental follies one or another of these grand designs for reconstructing himself, man would then show sudden increase in wisdom. If genetic policy-making were not miraculously improved over public policy-making in environmental and political matters, then access to the Tree of Life (meaning genetic management of future generations) could cause grave damage. It could cause the genetic death God once promised and by his mercy withheld so that his creature, despite having sought to lay hold of godhood, might still live and perform a limited, creaturely service of life. Then would *boundless freedom* and self-determination become *boundless destruction* in its end results, even as its methods all along included the unlimited subjugation of man to his own rational designs and designers. No man or collection of men is likely to have the wisdom to rule the future in any such way.

Genetic Responsibility, Genetic Treatment

Nothing said above should inhibit the use of genetic information in the practice of medicine, nor is it intended to do so. In medical practice the patient is an individual human being and his or her probable offspring. The patient is not "society" or the "future" or distant "generations" as such. Having rejected on moral and as well on scientific grounds some of the grand designs for treating or improving these nonpatients, it might be well in conclusion to summarize some of the things we should now be doing in actual medical practice and in public policy related to it. From these primary concerns we ought not to be distracted by an overweening desire to master the future of our species.

There is, of course, no question about the need for more extensive genetic counselling, and for the education of people about the availability of these services. As our knowledge of inherited grave defects becomes more exact, and especially when geneticists are able to identify more of these diseases before one instance has occurred in a family, there will be need for premarital genetic counselling. If we require premarital blood tests to protect the innocent or the unknowing from being infected with venereal disease, it would be entirely proper as a matter of public policy (if and when the science is exact enough) to require certain premarital genetic tests to protect the innocent and the unborn, and the unknowing partners contemplating marriage, from complicity in a tragic birth they may not and should not want.

We could as a matter of public policy, and probably should, go further. There may well be defects that are identifiable—say, in the case of grave dominantly inherited diseases— which would warrant using the marriage licensing power of the state to prevent their transmission. This would be in the service of life and in the service of the only patients known to medical practice. After all, it ought never to be believed that everyone has an unqualified right to have children, or that children are simply for one's own fruition. Instead, *pro*-creation is the service of human beings to come, and the ordinary right to have children could readily become in given instances the duty *not* to do so. Then, beyond premarital tests and conselling, it is arguable that it would in certain cases be desirable for public policy to condition and restrict marriage licenses by this present or future knowledge in genetics. This should not be taken *ab initio* to be too great an infringement of human liberty, because the freedom of parenthood is a freedom to good parentage, and not a license to produce seriously defective individuals to bear their own burdens.

Anyone concerned with genetic counselling and the practice of genetic medicine in the service of life should, then, be set against all those influences in our society leading to steadily increasing illegitimacy and to the mores by which getting pregnant is fast becoming an accepted way, or an "institution," for getting married. These practices are sufficient to void the individuals' voluntary genetic responsibility and the premarital tests or conditions that might be warranted in a rational use of our growing knowledge of genetics. We need to mean what we say when we speak of responsible parenthood.

Moreover, sexual intercourse and/or marriage at an increasingly early age are known to be dysgenic. It is well known that young women no less than women who bear children past the age of thirty-five have a greater incidence of chromosome difficulties. Studies made in Japan where the average upper age of childbearing has rapidly gone down to thirty-five, while the age when women have their first child has over recent decades gone up to twenty, have shown not only a decrease of mongolism, which rises exponentially when the woman is of older age, but also a marked decrease of congenital defects attributable to childbearing at an early age.[22]

As a final illustration, there are abundant studies which show that, in mammals other than man, sperm ejaculated in intercourse some time before the ovulation of the eggs they fertilize are apt to produce a far greater

incidence of congenital malformations than arise from
sperm ejaculated closer to the time of ovulation. So
there are scientists who believe that in human beings the
practice of the rhythm method of conception control is
not wrong because it doesn't work, but wrong because it
does! That is to say, the essence of the rhythm method
is to locate sexual intercourse as far as possible from the
time of ovulation. If, nevertheless, conception takes
place, it will be from an older sperm than might otherwise be the case, and not from a random sample of these
"time-acts" of intercourse. The studies mentioned above
seem to show that this is dysgenic. Some scientists,
therefore, believe that the time may come when we have
enough evidence to know that we should all practice
contraception for all the time considered infertile by the
rhythm calculations, and should have procreative intercourse only when the temperature chart goes up, in
order to *optimize the conceptus* from sperm as fresh as
the ovum it fertilizes. Even as now the temperature
chart is used to enable a woman to have a child, it could
be used as a way of enabling women to avoid a significantly greater incidence of congenital deformities.[2][3]

With all these "sooner anxieties" and the very great
many things to do in the application of genetics in the
service of life, it would seem prodigal to devote thought,
energy, and educational resources to the practice of
genetics on that nonpatient: the species.

In general, grand designs of positive or progressive eugenics should be opposed in the name of the right and
the good, in the name of more urgent and practical
applications of genetics in medicine—and because of our
lack of wisdom to create, by favoring selected phenotypes, an evidently stronger and more adaptable species
than nature has achieved in man pluralistically.

But what of *negative* eugenics, introducing chemical changes in the genotype (genetic engineering)? What of genetic surgery for the sake of the child to be conceived? If in future acts of genetic surgery we stick close to assured knowledge of the gene that needs to be rendered inoperative, and do not blend this with grandiose notions of genetic improvement of the species, a preventive use of genetic engineering in behalf of the child about to be born cannot be denied a possible place in genetic medical practice in the service of patients. The present writer has yet to discover any reason to reverse the *formal* judgment made upon first venturing into this science-based moral question: "Should the practice of such medical genetics become feasible at some time in the future, it will raise no moral question at all—or at least none that are not already present in the practice of medicine generally. Morally, genetic medicine enabling a man and a woman to engender a child without some defective gene they carry would seem to be as permissible as treatment to cure infertility. . . . Any significant difference arises from the vastly greater complexity of the practice of genetic surgery and the seriousness of the consequences if, because of insufficient knowledge, an error is made. . . . The science of genetics (and medical practice based on it) would be obliged both to be fully informed of the facts and to have a reasonable and well-examined expectation of doing more good than harm [to *this* patient, this progeny] by eliminating the genetic defect in question" (see above, pp. 44-45).

Joshua Lederberg's writings have made me more aware of the danger of "slipping a single nucleotide" in genetic surgery. This, and the complexity of our genetic mechanism in general, might mean that it is unlikely that we will ever know enough assuredly to control remedial medical applications of genetic surgery—when this surgery is submitted to the test of "a reasonable and well-examined expectation of doing more good than harm" to the patient to be born. I still see no reason for removing that *formal* ethical verdict upon this possibility. *Materially,* that formal verdict may have to be reversed. That is to say, when we come to an actual comparison of the good to be done or defect to be prevented in the patient to be born with the likelihood of as great or greater damage, a negative answer may have to be given to the proposal of genetic surgery. In a straight comparison, obviously, a hazardous surgical trial is to be preferred to grave genetic defect. But in a comparison of two great and likely evils, one must ask whether the choice is disjunctive, whether either must be chosen, whether there is not a third alternative. This is particularly true of the first experimental trials of genetic surgery on human beings—to get to know how to do it and to perfect the therapy. Can *this* morally be done? is the crucial question (see below, pp. 116-20).

In any case, applications of our genetic information in medical practice should never lose sight of the physician's real patients: the man, the woman, and the child to be born. Nor should they replace them by nonpatients: the species, or our control of human evolution. If these applications include preventive genetic surgery, it is because such action is required and reasonably may be done in the service of these lives.

It is significant that when proponents of positive eugenic designs are hard pressed, they frequently repair again to this sound limit upon the practice of medicine upon human beings. Dr. Leon R. Kass, a biochemist at the National Institutes of Health in Bethesda, Maryland, responded to Lederberg's column in *The Washington Post* (Sept. 30, 1967), in a letter to the editor.[24] He pointed to the fact that scientific and technical progress has a momentum of its own which is apt to erase the distinction between "it can be done" and "it should be done" or "it has to be done." "Biologists today," he wrote, "are under strong obligation to raise just such questions publically so that we may deliberate *before* the new biomedical technology is an accomplished fact." He asked: "Is the human will sufficient authority to advocate or to attempt to clone a man?"; "Should an independent scientist carry out such an experiment in the absence of public authorization?"; "Who should control the genetic planning?"; In case of mishaps, "who will or should care for 'it' and what rights will 'it' have?"; "Will the programed reproduction of man not in fact de-humanize him?"

Dr. Lederberg responded in his column in the *Post*;[25] his reply focused unduly upon the expression "tampering with human genes." He stated that there would be "no greater difficulty about reversing exolutionary mistakes than there were in making them in the first place" —as if that constituted an answer to the serious issues Dr. Kass raised.

Then came a surprising paragraph, which in the light of our foregoing discussion the reader may, not without reason, regard as disingenuous and less than forthright. I cite it now because in it Dr. Lederberg repaired to preventive eugenics and to a quite proper practice of genetics in medical service to the lives of people: "In fact, the anxiety about genetic intervention is almost certainly not directed to anything really likely to happen. We probably will welcome any chance to alleviate the impact of mongolism, schizophrenia, diabetes or dwarfism.

Probably what is more alarming is the abstract concept that 'man will control his own destiny.' Man as manipulator is too much of a god; as object, too much of a machine."

This strategic retreat happened again in the testimony of Drs. Kornberg and Lederberg before the Senate Subcommittee on Government Research on the Mondale bill, according to one newspaper account. Senator Ribicoff used scare-language in his question: "Do you see this leading to a master race?" No, at least probably not, was the answer. Then, instead of a frank discussion of what it is believed will be going on in laboratories at a very early date, both scientists said they saw this as only a remote possibility—unless, said Dr. Kornberg, by "master race, you mean *healthier* people" (italics added).[26] This—not doctoring that nonpatient, the human race—is the object of medical care.

Chapter 3: Parenthood and the Future of Man by Artificial Donor Insemination, Etcetera, Etcetera

> I have a pet hen whose name is
> Probable. She lays eggs in concept,
> being a sophist-bird. But not in
> reality at all; those would be
> inferior eggs; for thought is superior
> to reality.
>> Frederick Winsor, *The Space Child's Mother Goose*

Aldous Huxley's fertilizing and decanting rooms in the Central London Hatchery (*Brave New World*) will become a possibility within the next fifteen to fifty years. I have no doubt they will become actualities—at least as a minority practice in our society. One reason this will come to pass is that philosophers, theologians and moralists, churches and synagogues, do not have the persuasive power to prevent the widespread social acceptance of morally objectionable technological "achievements" if they occur. Philosophers whose business it is to transmit wisdom which begins in wonder and theologians whose business it is to transmit wisdom which begins in fear of the Lord,[1] while criticizing, reshaping, and enlivening these wisdoms, have collectively abandoned understanding, and their voices. This may seem a harsh and despairing charge. Perhaps the dehumanizing tendencies of technology in all advanced societies should simply be described as irresistible. But, if so, a first evidence of that irresistibility is the way in which leading intelligences, including theologians and churchmen, rush to offer the sacrament of confirmation or to celebrate a Bar Mitzvah before the event whenever they hear of any new means by which man will become a "self-modifying system"—his own creator, the unlimited lord of the future.

The Fascinating Prospect of Man's Limitless Self-Modification

There are profound anthropological and ethical issues raised by the possible future technical biological control and change of the human species (just as there are profound anthropological and ethical issues raised by the challenge to individual human self-awareness by the prospect of keeping alive a wholly "spare-parts" or an "artificial" man). To follow out either of these directions and long-range consequences of present research and development (which originates, to be sure, in a legitimate concern for the treatment of present human ills) would be to take a larger overview than in either previous chapter.

Physicians generally are content to stick close to present and near-future patient care—waiting until the longer day's dawning to consider whether some of the notorious things now proposed and only remotely possible should ever be done. Biochemists and molecular biologists, however, are keenly aware that research and development in the self-modification of the human species cuts all questions loose from the moorings of an ethics of medical practice and from the ethics upon which our civilization has so far been founded.

The fascinating prospect of man's limitless self-modification is almost daily placed before the public in magazine and news articles. With these prospects we must deal, and at least make the attempt to articulate the elements of a possible line of moral reasoning concerning them.

I intend, therefore, to draw together some of the themes introduced by the preceding two chapters and to set them in the perspective of additional possibilities ahead. Thus we shall see what may happen to medical ethics, or to ethics in general, when the future of the species is taken to be a patient who is to be reworked by biological technology and through new forms of human "reproduction."

One journal article, by David M. Rorvik, in the April 1969 issue of *Esquire*, was accompanied by appropriate pictures of specially bred astronauts, legless for efficiency on long space-voyages; a completely germ-free human being for colonizing outer space; short-legged stocky dwarfs for planets with high gravitational pressure; four-legged human types for Jupiter; men with prehensile feet and tails to hang on to planets with low gravitational pull; clones of Barbra Streisands, Mahalia Jacksons, Joe Namaths, and Adolph Hitlers to entertain us here on earth; chimeras[2] and cybogs to do janitors' work. This, of course, was sensational journalism. The wonder is that there is no outcry.

At the same time, the *Esquire* article printed the Rand Corporation's table of human expectations: artificial inovulation in humans by 1972; genetic surgery by 1995; routine animal cloning by 2005; widespread human cloning by 2020; routine breeding of hybrids and specialized human mutants by 2025. A.D., that is! And there was good, if sparse, scientific information contained in the article. This included accounts of: the work of the French scientist, Jean Rostand, in parthenogenesis by jolting frogs' eggs; H. J. Muller's campaign for germinal choice from banks; the experiments of Dr. J. K. Sherman of the University of Arkansas in successfully impregnating women with sperm from stocks frozen for prolonged periods at 385° below zero;[3] Dr. Sophia Kleegman's practice of artificial insemination from anonymous donors (AID) at $25 per masturbation; the transportation of an entire "herd of prize sheep" by air from Europe in the form of tiny embryos kept alive in the uterus of a single rabbit and then implanted in ewes—showing what can soon be done by inovulating humans; the way in which human motherhood could be made obsolete by ectogenesis—a combination of test-tube babies and the artificial placenta;[4] the work of Dr. O. S. Heynes of South Africa in putting women during the last stage of pregnancy in a special decompression chamber to increase the flow of oxygen to the fetus so as to produce more intelligent children;[5] how to avoid the limitations of the human female pelvis so the human brain can grow bigger; the laser beam that will make genetic manipulation possible; and, finally, the major "evolutionary perturbation" of clonal reproduction which men can seize for their profitable self-modification as a species.

Artificial Insemination with Donor (AID) is only the first breach of what has until recently been understood to be human parenthood as a basic form of humanity. Then there is artificial inovulation. And after artificial inovulation comes "germinal selection" from ovum and semen banks for the management and self-modification of the future of our species. Then, electro-tickle parthenogenesis, whose result would be only women, men no longer being needed.[6] Then, women hiring mercenaries to bear their children, as now they secure the cooperation of semen donors. "Sooner or later," writes Dr. Roderic Gorney of the U.C.L.A. School of Medicine, "a patient will request and get artificial gestation for her baby just because she is tired of the restrictions of pregnancy and wants to take a round-the-world tour or go skiing." Or there will be a woman offering to give gestation to the child of a dying sister who wants before passing away to leave her husband a child.[7] There will be: babies produced by reworking male and female germinal material in hatcheries, which unfortunately would at first still require somebody's womb to bring the fetus to term; the making of "carbon-copies" of people by clonal reproduction (using nuclear transplantation); clonal farming, offering everyone who can afford it a supply of "identical twin" organs whenever he needs a transplant;[8] the manufacture of short-legged astronauts or of a race of serfs by combining human with animal chromosomal material; the predetermination of the sex of our children, involving the zygotic, embryonic or fetal destruction of the unwanted sort. Compulsory, or at least the socially sanctioned, injection of young women with long-time contraceptives to enable them to maneuver through their early years without pregnancy (this might be a condition for admission to college).

The latest article to appear on these apparently fascinating prospects is "The Second Genesis" by the distinguished science writer, Albert Rosenfeld, published in *Life* magazine in the June 13, 1969 issue.[9] Beneath a sense of man's boundless freedom Rosenfeld's writing is suffused by a sense of man's boundless determinism: man the self-creator seems also the slave of the actions that biology now makes possible. The control of life is "coming," according to the book's subtitle. The miracle worker is bound to do these actions, because he can, or someone will; and the doer of miracles is destined to be indefinitely reshaped by performances to which he seems drawn out of dizziness before the prospects.

Rosenfeld covers the ground we have already tra-

versed—"solitary generation" (Rostand's phrase); artificial androgenesis (producing all males, as parthenogenesis would produce all females); banks of ova and
sperm; propagation by cuttings; sex selection by dia-
phragms to separate androsperm from gynosperm; *in
vitro* babies ("too great to resist"); hybridization with
animals; offspring not of particular couples even in culture but of the entire species or entirely fabricated
according to specification; the peddling of Celebrity
Seed; women competing with women not for marriage
partners but for sex partners while more and more men
cop out because of the celebrated female orgasmic
powers; wholesale automanipulation; and Dr. E. S. E.
Hafez's projects for combining AID with the production
of centuplets by induced superovulation in women, to
secure a supermarket of embryos for use on earth and to
miniaturize the people to be sent aloft to colonize the
planets.

It is evident that women who are to be freed by
"labor saving devices" are also to be limitlessly used.
This is only a special case of the "limitless freedom/lim-
itless submission" (Dostoevski) which will result with
the destruction of parenthood as a basic form of humanity, and its recombination in various ways for extrinsic
purposes.

In the end, Rosenfeld remarks:

> In our current circumstances, the absence of a loved
> one saddens us, and death brings terrible grief. Think
> how easily the tears could be wiped away if there were
> no single "loved one" to miss that much—or if that
> loved one were readily replaceable by any of several
> others.
>
> Ane yet—if you (the hypothetical *in vitro* man) did

not miss anyone very much, neither would anyone miss *you* very much. Your absence would cause little sadness, your death little grief. You too would be readily replaceable. . . .

The aloneness many of us feel on this earth is assuaged, more or less effectively, by the relationships we have with other human beings. . . . These relationships are not always as deep or as abiding as we would like them to be, and communication is often distressingly difficult. Yet . . . there is always the hope that each man and woman who has not found such relationships will eventually find them. But in the *in vitro* world, or in the tissue-culture world, even the hope of deep, abiding relationships might be hard to sustain. Could society devise adequate substitutes? If each of us is "forever a stranger and alone" here and now, how much more strange, how much more alone, would one feel in a world where we belong to no one, and no one belongs to us. Could the trans-humans of post-civilization survive without love as we have known it in the institutions of marriage and family?[10]

The American public, when questioned about their approval or disapproval of these ways of "improving" human reproduction, manifest a surprising degree of approval of them. A Harris Poll published in *Life* magazine, on June 13, 1969, showed that two people disapproved for every one who approved of AID, artificial inovulation and *in vitro* babies. But the fact that one out of three approved was the remarkable thing. It is notable, however, that people's approval is based on interpreting these proposals as treatments. They do not have in mind the primary purpose and effects that are often in view in the case of some scientists. While fifty-six percent disapproved and nineteen percent approved of AID upon a simple description of the procedure, thirty-five percent approved when this was explained as the only way by which a couple could have a child or a normal child. By contrast, a larger number, thirty-six percent of both men and women, surprisingly approved of artificial inovulation if the husband fertilized the egg. The explanation was that both the men and the women believed that men feel emasculated by AID. *In vitro* babies won the approval of twenty-five percent of both men and women, while thirty percent of the men and thirty-five percent of the women approved if the wife might die or be crippled from childbirth. The hazards to babies in learning how to achieve this were not brought into consideration by the pollsters. Purposely making genetic changes in order to "give a child to an otherwise childless marriage" won sixty-two percent approval; to avoid retarded babies won fifty-eight percent approval; while producing superior people through genetics was roundly rejected by fifty-seven percent to twenty-one percent. Again, the hazards to the unborn child were not reported to have been brought into question.

The conclusion to be drawn from the Harris Poll is that approval of such novel proposals is related to the treatment of infertility in marriage, the prevention of birth defects, and the preservation of man's sense of wholeness. The remarkable thing was not the degree of acceptance but that the acceptance was based on care of the persons involved in the family or of one another. The poll also disclosed a great amount of pro-natalist sentiment and the belief that people must have children.

Little comfort should be drawn from these observations, however, if the future development of people's acceptance and future behavior patterns in our society show that treatments offered in primary patients, people's present care of one another, and even their pronatalism *by any means* may be readily misused and molded into self-modifications of the transmission of life as a means to other goals. This could be the inexorable result of our present fascination with biological techniques applied to the origins of life.

How Far Are These Procedures Legitimate Treatments?

I am aware that many of the researchers and practitioners who are developing these exquisite "remedies" are motivated primarily by *therapeutic* goals. Several examples of such good uses of the relevant scientific know-how can be given.

Artificial insemination was first developed to enable a husband and wife to have children. Making it possible for a woman to have a child is the purpose even when nonhusband donors are used. The practice need not be directed to the improvement of that celebrated nonpatient, the human species.

Artificial inovulation also has a therapeutic purpose and effect. If a woman's ovum for some reason cannot reach the place of impregnation or be impregnated by her husband, it is possible for the physician to extract it, fertilize it with semen from her husband, and then implant it in her uterus. This assists fertility in the marriage. Such an excellent treatment, however, becomes at once a possible way of circumventing the normal parameters of parenthood, and a possible way of treating future multitudes.

Similarly, caesarian sections constitute descriptively and ethically one sort of procedure when it is indicated that normal birth would be unsafe for both mother and child. But the use of caesarian sections routinely on all women who have babies in hospitals in order to overcome the restrictions of the human pelvis and let the human brain grow larger and larger over generations to come (suggested by Joshua Lederberg)—that, morally, would be altogether a different procedure. It would introduce unknown evolutionary perturbations and use today's women for the purpose of doctoring the species in a later time.

Again, the improvement of present incubator methods—an artificial placenta, no less—might have great value in saving the lives of the prematurely born. This cannot be regarded, however, as a procedure to be chosen in place of the fetus' being nine months in the womb of its mother, when one remembers the disadvantages of prematurity—ranging from greater mortality to serious mental and physical impairments in development.

Let us imagine that there can be developed an artificial placenta as good for the child as the womb—or better, because it abolishes the limits imposed by the human pelvis upon brain development, and makes the child accessible to "the management's" improvement. Even so, such a technical development skips over the crucial ethical question. Prescinding from the "good" ends in view, the decisive moral verdict must be that we cannot rightfully *get to know* how to do this without conducting unethical experiments upon the unborn who must be the "mishaps" (the dead and the retarded ones) through whom we learn how. It is amazing that, in discussions of man's self-modification of the future of his species by prenatal refabrication, this simple, decisive ethical objection is so seldom mentioned. This can only mean that our ethos is well prepared to make human waste for the sake of these self-elected goals.

Intrauterine monitoring for the purpose of detecting in the unborn genetic or developmental defects which may be corrected by fetal surgery or other procedure is an obvious accomplishment in extending treatment at a time when it might be beneficial—treatment to which physicians formerly did not have access. (Whether by these extraordinary means unborn lives should always be saved is another question.) But at least some forms of "screening" (in contrast to monitoring and treatment) focus upon patients other than the unborn, and propose the elimination of some genetic defect from the species by eliminating these primary patients after they are here among us as lives yet unborn—including those who are only carriers and would themselves be quite healthy. The public is dimly aware of the fact that abortion as a means of population control is only a form of increasing the death rate (where, as in our age, the control of death has helped to create the population problem). But still fewer people are aware that abortion—even abortion on the so-called "fetal indications" of a probability of grave mental or physical defect in the child—can readily cease to be even a form of alleged therapy for that child.[11] "The abortion dilemma," it has recently been pointed out, "is only the currently visible small fraction of the very large iceberg dealing with the control of the quality of human life,"[12] which is the generalized "patient" to be dealt with by these abortions and other procedures. Abortion will soon become a way of doctoring that non-patient: the species.

Amniocentesis to detect congenital defects (and if the defects are sex-related, the sex of the child) has therapeutic value where there is some present or future method of extending care to the patient. The proposed use of this procedure, however, or the use of artificial inovulation, to predetermine the sex of the child (in the latter case, by not implanting the undesired sort) departs from the parameters of human parenthood so far as to raise the gravest social and ethical questions.

In our present society, for example, this would likely lead to a serious imbalance between the sexes of the children chosen to be born. In spite of what we say about the equality of the sexes, and about liking girl-children as much as boys, our operative decisions would be to the contrary. Statistics plainly show that parents seek to have an additional child when they already have one, two, three, or four girl-children far more frequently than they do when—in each of these classifications—they already have one, two, three, or four boys. Generally, people often try again for a boy, less often for a girl. Professor Amitai Etzioni, sociologist at Columbia University, has pointed out the widespread social, moral and political repercussions—including a return to the frontier atmosphere in this country—that would follow any attempt to predetermine the sex of our children.[13]

Finally, genetic surgery by means of laser beams or some chemical to reverse mutations might be a wholly acceptable procedure for treating the primary patient—to correct some serious genetic defect with which a child is otherwise going to be born. In this case, any mishap resulting from the process of trying to knock out a nucleotide or change the child's genetic makeup would be a great tragedy; but it would be a consequence of decisions and actions taken in behalf of the child's health. This must be classed among the normal hazards of proper medical care. If there is some miscarriage, it is not a miscarriage of justice—as would be the case if the mishap resulted from experimenting on the child in a program of positive eugenics for the supposed sake of the species.

Since the foregoing point repeats the judgment I have made elsewhere (see above, pp. 44-45, 101) concerning the treatment-value of genetic surgery—perhaps the most exquisite of all the procedures we have under review—let us look more closely at the question of whether this is a *choice-worthy* treatment and at the ethically relevant circumstances that *could moraliy deny us this option.*

Soon laboratories will be submitting requests for federal research funds to finance the search for viruses to manipulate defective genes. The alteration of genes by viruses will, of course, be done first in animals. If promising viruses are discovered, application must then be made to the Food and Drug Administration to license the material for testing in humans. This brings us to the crucial point of whether our knowledge at that stage will be such that any mishap can correctly be classed among tragedies in the practice of medicine—this time, upon a hypothetical human being, the unconceived child[14]—or whether the hazards will be such that it would be immoral to proceed further with *the attempt to learn how* to use viruses to change genes.

As I understand it, one gene must be replaced by another; to manipulate out a bad gene entails the introduction of another genetic determiner. It may be possible to find a virus that would carry only the desired gene, or at least one that would be known to carry no deleterious genes. In that case the moral objection about to be raised would not pertain. However, the biophysicist, Leroy G. Augenstein, of Michigan State University, describes the situation we might face in deciding whether to begin testing genetic surgery on human beings. "Suppose we were to find a virus which carried the necessary DNA for correcting diabetes and made all the boys very tall (good basketball teams) and raised their IQ's by fifteen points (no flunking out of school)." Then if anything went wrong, it would be tragic, not an immoral act we had done. Suppose, however, Augenstein continues, "we were unlucky and the virus contained not only a certain amount of DNA enabling people to make their own insulin, but additional DNA so that the group tested either went on to have defective children or developed schizophrenia. We would have a whole generation with extensive genetic changes *before we even knew they were in trouble.*"[15]

Obviously, if we knew beforehand that these would be the results, the introduction of a virus to correct the insulin production of an as yet unconceived child would be no proper treatment. It would rather be a wicked thing to do. But suppose we do not know that these terrible consequences of tampering with the gene for insulin will be forthcoming? What then? Is it only tragedy as a hazard of proper medical care if and when these things result? I should say not. Given the intricate and wonderful structure of the genes and the lottery that produces the genotype which is or becomes a human being, we ought morally to require a far higher degree of knowledge that *there are no hazards* of such gravity. It is not enough not to *know*, one must rather *know* that there are *not* these hazards, before this homing-in on the gene for insulin could possibly be a choice-worthy treatment of a hypothetical human being, as yet unconceived, who seems likely to become diabetic.

This is not only because insulin injections are better treatments. Suppose it were judged that diabetes is serious enough for the individual and that the increasing number of diabetics arising from the gene pool is so serious that something must be done about it. Still, we ought not to choose genetic surgery at the risk of producing individuals who may, in increased numbers, become schizophrenic in the first generation and who may be the mothers and fathers of children who are defective in the second generation, even if their defects are deemed to be no greater than diabetes. The reason for this conclusion is that there is a third alternative for treating the unborn. We are not forced to choose between doing nothing about diabetes as an inherited disease and correcting it by genetic manipulation under these supposed conditions. The indicated treatment for preventing the transmission of diabetes would be having no children or fewer children. The treatment would be continence or not getting married or using three contraceptives at the same time or voluntary sterilization. Only someone who is more of a pro-natalist than the Roman Catholic Church ever was, or who strictly believes that every human being has an absolute right to have children, can avoid the conclusion that these are more choice-worthy options than visiting upon hypothetical children the risks that we are supposing are associated with the removal of diabetes, or that we must suppose *have not been removed* from the realm of possibility. The treatment of choice would be to do everything possible to correct the consequences of "achievements" of the past, when from an unlimited and unexamined pro-natalism we learned how to enable diabetic women to have children.

Let us make the case a harder one, to see if these ethical conclusions do or do not still hold true for the meaning of genetic responsibility. Suppose the bad gene believed to be manipulable by a virus is far, far more serious than diabetes, e.g., a recessive trait like cystic fibrosis or PKU; or a dominant trait like achondroplasia (dwarfism) or Huntington's chorea where the defect will be passed on with a fifty-fifty probability (like brown hair or brown eyes). What, we now ask, ought rightfully to be done in behalf of, first, the child prenatally and, secondly, the hypothetical child preconceptually?

If amniocentesis, intrauterine monitoring, etc., disclose the fact that an unborn child has two doses of the recessive genes for a serious illness or has proved unlucky on the fifty-fifty chances that he will be a victim of one of these serious dominant defects, without question his parents can rightfully consent to drastic prenatal treatments—since these treatments may, it is reasonable to believe, be beneficial to him. Then if the worst befall, it would be the tragic result of rightful actions. We could go further and say that even if the unborn child is not certainly afflicted with one of these diseases, he may be one of a limited population at grave risk. (Amniocentesis, etc., backed up by a reading of the unborn child's maternal and paternal genetic history can indicate this risk.) We could say that parents can validly consent in behalf of such an unborn child medically, permitting the physician to use possibly beneficial trial treatments that, however drastic, do not place the child at greater risk than now surrounds him as one of a specially endangered population.[16]

It is obvious, however, that canons of loyalty involved in the treatment of an as yet unconceived life—canons of loyalty similar in any way to those that are standard for our treatment of an actual child prenatally or postnatally—will require far more certainty of possible overall benefit to the hypothetical child before genetic surgery could be the treatment of choice. There would have to be far more than a fifty-fifty probability that a hypothetical child relieved of dwarfism would not be afflicted with schizophrenia (or would not produce children with other defects in the second generation), or that the child would not suffer worse mishap if relieved of the statistical likelihood that he will suffer Huntington's chorea or cystic fibrosis.

This way of expressing the option is, in fact, quite false, since we are not forced to choose between genetic surgery and doing nothing at all to prevent the conception and birth of these children. A hypothetical child is nothing—or at least nothing until we begin working on him preconceptually upon the fixed assumption that he will be engendered. Therefore, there would have to be at least *no discernible risks* before genetic surgery would be for him the treatment of choice. Before a child is at all actual he has no title to be born. Men and women have no unqualified right to have children. The treatments for the prevention of cystic fibrosis, Huntington's chorea, achondroplasia, some forms of muscular dystrophy, PKU, amaurotic idiocy, and other chromosomal abnormalities (if and when our Early Genetic Warning System can be perfected to detect them before conception) are continence, not getting married to a particular person, not having any children, using three contraceptives at once, or sterilization.

In discussing the ethics of the crucial step that begins the trial of genetic surgery upon humans, we should demand to know why these alternatives are not genetically more responsible. Finally, we must observe with some amazement that we live in an age that can calmly contemplate these two contradictory procedures: (1) abortion when there is likelihood that the child will be seriously impaired mentally or physically, and (2) learning how to do genetic surgery on humans although this may lead to the conception of children who may be seriously impaired mentally or physically (the mishaps).

It may be unfair to attribute to geneticists, who write as if they are not to be deterred by a proper ethics of treating hypothetical children, the pro-natalist attitudes of past traditional societies. If not, the explanation of their easy assumption that genetic surgery is a procedure which, when it becomes feasible, should be put into actual use may be that for them "genetic manipulation is only the currently visible small fraction of the very large iceberg dealing with the control of quality of human life" generally, having in view man's improving self-modification. A subtle but significant shift has taken place from doctoring primary patients to doctoring that nonpatient, the human race. For this reason, patients now alive or in the first and second generation may be passed over lightly, hypothetical children can be thought of as actualities to be improved at risk, and one can even contemplate permitting harm to come to them (with abortion as an escape prepared for the injured) for the sake of knowledge and learning the techniques ordered to the good to come.

While in the foregoing I have unavoidably stepped upon the terrain of ethical reasoning, my chief intention has been to show that actions whose objective is treatment and actions whose objective is the control of the future of our species are different sorts of actions, even when descriptively they may look alike; and that these two different sorts of actions *may* be subject to opposed ethical limits or evaluations.

The Ethical Questions

"If a scientist fixated upon the technical difficulties of

the feat produces a clonal offspring cultivated from some medical student's intestinal cells, no judge can then decide, when someone comes tardily to court, that the baby should be uncreated."[17] The same can be said of the other feats we have reviewed. If a mishap from trying genetic surgery upon humans comes tardily to court, the judge cannot say that baby should be uncreated. If an embryo created and nurtured to development *in vitro* comes tardily to the court of public opinion, we cannot say that baby should be uncreated.[18] It will be too late to say it ought not to have been created. Discussion of the moral questions raised by the new biology must begin now. In this discussion the public should be engaged, and from it no helpful perspective should be excluded. The humanity of man is at stake. In ensuring that man shall remain man when and after he does any of these projected procedures for the increase of knowledge and for his own improvement, consequences are an important consideration. In fact there is nothing more important in the whole of ethics than the consequences for good or ill of man's actions and abstentions—*except* right relations among men, justice, and fidelity one with another. The moral quality of our actions and abstentions are determined both by the consequences for all men and by keeping covenant man with man.

We need to raise the ethical questions with a serious and not a frivolous conscience. A man of frivolous conscience announces that there are ethical quandaries ahead that we must urgently consider before the future catches up with us. By this he often means that we need to devise a new ethics that will provide the rationalization for doing in the future what men are bound to do because of new actions and interventions science will have made possible.[19] In contrast, a man of serious conscience means to say in raising urgent ethical questions that there may be some things that men should never do. The good things that men do can be made complete only by the things they refuse to do.

Unavoidably, in outlining and analyzing the prospects before us, we already have introduced ethical considerations. I propose in the following sections of this chapter to raise four questions that seem to me crucial in making a proper response to the issues raised by the new biology (only the first of which has strictly to do with the consequences for good or ill of the adoption of one or another of these proposals for doctoring the species): (1) the question of whether or not man has or can reasonably be expected to have the wisdom to become his own creator, the unlimited lord of the future; (2) the anthropological and basic ethical question concerning the nature and meaning of human parenthood, and of actions that would be destructive of parenthood as a basic form of humanity; (3) the questionableness of actions and interventions that are consciously set within the context of aspirations to godhood; and (4) the question of human species-suicide. Some of these topics will be discussed more fully than others, and it is quite impossible (as we shall see from the very beginning) to keep them separate from one another. Questions of ethics are from the beginning questions of philosophy, of total worldview, of metaphysical or ultimately religious outlooks and "on-looks." It may be helpful to bring these things to the surface, in an age when many men imagine there can be ethics without ultimates.

A Question of Wisdom

First, then, the question of whether man is or will ever

be wise enough to make himself a successful self-modifying system or wise enough to begin doctoring the
species. When concern for the species replaces care for
the primary patient, and means are adopted that are
deep invasions of the parameters of human parenthood
as it came to us from the Creator, will we not be
launched on a sea of uncertainty where lack of wisdom
may introduce mistakes that are uncontrollable and irreversible?

To this it is no answer to say that changes are already
taking place in human kind, or that men are constantly
modifying themselves by changes now consciously or
unconsciously introduced, for example, in the environment. It is no answer to say simply that in the future
proposed to us by the revolutionary biologists the
changes we are now undergoing will only be accelerated
in rate or that the self-modifications then going on will
simply be deliberate in major ways. It is true that man
adapts to his environment and that his environmental
changes change him. The point now being made, however, may be cinched by saying that, from man's rape of
the earth and his folly in exercising stewardship over his
environment by divine commission, there can be derived
no reason to believe that he ought now to reach for
dominion over the modifications of his own species as
well. It is almost a complete answer to these revolutionary proposals simply to say that "to navigate by a
landmark tied to your own ship's head is ultimately
impossible."[20] Many or most of the proposals we are
examining are exercises in "What To Do When You
Don't Know The Names Of The Variables,"[21] not even
the variables which our beginning to act upon the proposals, or our making of the proposals, may bring to
pass in human society generally. The proposals are spec-

ulative speculations, not programs. They could not be otherwise. We do not even know how to learn to predict the consequences of presently projected lordships over the future in remote future human situations whose values and milieu we have no means of controlling. The following proposition is, therefore, as good as any other. Man cannot endure if there is no creation beneath him, assumed in his being, on which he ought not to lay his indefinitely tampering hands.

This is the Theory Jack built.

This is the Flaw
That lay in the Theory Jack built.

This is the Mummery
Hiding the Flaw
That lay in the Theory Jack built.

This is the Summary
Based on the Mummery
Hiding the Flaw
That lay in the Theory Jack built.

This is the Constant K
That saved the Summary
Based on the Mummery
Hiding the Flaw
That lay in the Theory Jack built.

This is the Erudite Verbal Haze
Cloaking Constant K
That saved the Summary
Based on the Mummery
Hiding the Flaw
That lay in the Theory Jack built.

This is the Turn of a Plausible Phrase
That thickened the Erudite Verbal Haze
Cloaking Constant K
That saved the Summary
Based on the Mummery
Hiding the Flaw
That lay in the Theory Jack built.

This is the Chaotic Confusion and Bluff
That hung on the Turn of a Plausible Phrase
That thickened the Erudite Verbal Haze
Cloaking Constant K
That saved the Summary
Based on the Mummery
Hiding the Flaw
That lay in the Theory Jack Built.

This is the Cybernetics and Stuff
That covered Chaotic Confusion and Bluff
That hung on the Turn of a Plausible Phrase
And thickened the Erudite Verbal Haze
Cloaking Constant K
That saved the Summary
Based on the Mummery
Hiding the Flaw
That lay in the Theory Jack built.

This is the Button to Start the Machine
To make with the Cybernetics and Stuff
To cover Chaotic Confusion and Bluff
That hung on the Turn of a Plausible Phrase
And thickened the Erudite Verbal Haze
Cloaking Constant K
That saved the Summary
Based on the Mummery

Hiding the Flaw
That lay in the Theory Jack built.

This is the Space Child with Brow Serene
Who pushed the Button to Start the Machine
That made with the Cybernetics and Stuff
Without Confusion, exposing the Bluff
That hung on the Turn of a Plausible Phrase
And, shredding the Erudite Verbal Haze
Cloaking Constant K
Wrecked the Summary
Based on the Mummery
Hiding the Flaw
And Demolished the Theory Jack built.[22]

There are bound to come macabre suggestions for how men and women can help to avoid genetic disasters. Some of these will be aimed at eliminating existing spontaneously occurring genetic defects. Others will be directed at preventing the foreseeable genetic effects of unwisely adopted *man-made* solutions to other problems (AID). Thus, Linus Pauling suggests that, since the test for the presence of the gene for sickle-cell anemia in heterozygotes is extremely simple, "there should be tatooed on the forehead of every young person a symbol showing possession of the sickle-cell gene [or other deleterious recessive gene, so that] two young people carrying the same seriously defective gene in single dose would recognize this situation at first sight, and would refrain from falling in love with one another."[23] That may well be a measure needed to facilitate genetically responsible decisions on the part of people contemplating marriage.

Oddly enough, however, much the same proposal was

made in that same journal as a way of avoiding possibly disastrous consequences of AID. Estimating that donor insemination has now produced one million babies, Dr. Roderic Gorney of the U.C.L.A. School of Medicine writes: "The ease and anonymity of therapeutic insemination increase the risk that children unknowingly born of the same father will marry and reproduce, with all the possibilities of increasing genetic assets and defects that go with in-breeding. We may some day be forced to disclose paternity to the child, or perhaps encourage the wearing and mutual checking of coded dogtags at first dates."[24] Geneticists generally believe that normal reproductive patterns make for variety and sustain the adaptability of the human species, while keeping unexpressed many bad recessive genes that would, by widespread in-breeding, come to be manifest in a larger number of afflicted individuals. Another solution to this man-made problem is the suggestion that mothers who have had AID performed on them should have a secret signal—e.g., some gesture of the hand—which another such mother would recognize. They could then attempt to take action, if it appeared their AID children were about to get romantically involved with one another. They could try to do away with the real, however small or remote, possibility that their original acts may have bad consequences in the second generation for the grandchildren they and the same anonymous donor might have in common.

How then is AID a responsible decision or action or practice, any more than marrying someone known to possess the same gravely deleterious recessive gene? It is no answer to all this to say that sexual relations outside of marriage are producing a like anonymous parentage, and the same possibility of in-breeding. One does not excuse one genetically irresponsible action by appealing to another. Besides, AID *adds* to the children of normal adultery. How then can it be thought responsible as a mode of procreation—except on the assumption that there is always a scientific solution of any problem created by previous scientific solutions *ad infinitum?* Obviously, some deleterious changes may be scientifically reversible, while others may not. Still others will be irreversible because men cannot make the changes for which they have lost the capacity to wish. Was a boundless voluntarism, brooding over a boundless determinism, the summary based on the mummery hiding the flaw that lay in the theory Jack built?

Revolutionary biologists and the general public usually assume that men may now have an obligation to bring about a given situation in the *remote* future. Such an assumption deduces moral claims after assuming the role of Providence over the future. The concept of any such moral claim is seriously questioned in a perceptive article by Professor Martin P. Golding of the Department of Philosophy, Columbia University.[25] Such a concept of our moral relation to the community of the remote future, Golding writes, is far from clear; "certainly less clear than our moral obligations to communities of the present." The more remote, "the less do we know what to desire for them." One would not only have to specify the qualities to be enhanced. He would also have to show "that their enhancement would truly be advantageous for the community for which they are to be enhanced."[26] This burden of proof has to be assumed if it is proposed that, for the sake of the remote future, present moral values or claims are to be overridden or changed. It has not been assumed because no such conclusion can be given sufficient support by rational argument. The promotion of certain qualities could easily "be undesirable unless we could also determine the conditions of life of the community of the future, including its values." No one knows, for example, whether an increase in the number of intelligent men would be a good thing unless he could guarantee a comparable increase in the number of altruists.[27] One, therefore, now knows that he does not and cannot know that an increase in intelligence would be desirable in the community of the remote future. Only God knows, or (if that is past tense) only God could know enough to hold the future in His hands. Is this the flaw that lay in the theory that Jack built for doctoring the future?

The Destruction of Human Parenthood

Regardless of the knowledge about consequences which many of these proposals would require—a wisdom demonstrably impossible for men to have now about the *remote* future—many of these proposals would irreversibly remove a basic form of humanity: the basis in our creation for the covenant of marriage and parenthood.

Concerning hatcheries (or, as the scientists say, the production and genetic improvement of human beings *in vitro*—in *glass,* in laboratories), someone in that day will readily write: "From the point of view of the fertilizer and the decantor alone, procreation is depersonalized; from the point of view of personally self-conscious men and women, however, it is only debiologized. When we take a comprehensive view of all this, it is clear that neither the biological nor the personal dimension is absent." So saying, he will gain the ear of the groundlings—creatures of this world come of age—who do not have the moral courage or the ethical concepts, the religious daring or the theological concepts, with which to radically challenge the basic assumptions of a technological civilization.

That allegation in defense of hatcheries will undeniably be true. It is possible to combine the biological and the personal in all sorts of ways once their fundamental unity in the nature of human parenthood has been broken. In that day no one will speak falsely who declares that procreation has only been debiologized, that the biological has simply been put under (where it allegedly belongs), and that on any comprehensive view of these matters both the personal and the biological are still present and cherished.

Meanwhile, some theologians, I grant, are bothersome fellows. Especially so are theologians willing to explore the whole range of Jewish and Christian ethical categories and to apply them to the moral decisions facing individuals and society today and in the future. Human parenthood is, in the language of Karl Barth, a basic form of humanity. To violate this is already dehumanizing, even if spiritualistic or personalistic or mentalistic categories are invoked to justify it. To reason from *the person* (arguments about "intentionality," "commitment" to nurture the child, and "love" between the partners in these technological enterprises) provides only a part of an ethical verdict upon the current proposals of the revolutionary biology, and it is probably not the most fitting or fruitful beginning. And "soul-body" dualism is a downright error.

The parameters of human life, which science and medicine should serve and not violate, are grounded in the man of flesh and in the nature of human parenthood. Today anyone who affirms this will be accused of physicalism. Is there not, he will be asked, "a personal dimension" which is "something other" or "beyond" the "biological dimension"? From this small suggestion people are supposed to be convinced of quite another proposition, namely, that the biological "dimension" of his existence as a man of flesh really is an "other" entirely submissible to man's limitless *dominion.* This is enough to give unconditional baptism to any future medical or genetic technology, since from the means will come the goals they reach.

Any such view has lost all memory of the biblical view of man, and we may as well say so—techno-theologians to the contrary notwithstanding. Not that Athens has conquered Jerusalem, but wizardry has bemused both, a wizardry manifesting a radical dualism for soul (mind) and body.

The modern intellect swarms with more species of soul-body dualism than in any epoch of our historical past since Christianity replaced the decadent Neoplatonism of the Graeco-Roman world. It does not matter that in pre-Christian times the body was regarded as evil, or at least low, while in our post-Christian present age the body is believed to be good, or at least neutral. The "dualism" consists not in these value judgments, but in the "wholly otherness" between the self and its physiological life, energy, processes.

I am reminded of that strange translation in the Revised Standard Version of the New Testament (to which Roger Shinn recently drew my attention). There, where Jesus quotes Genesis on marriage: "The two shall become *one flesh,"* the RSV reads "the two shall become *one"* (Matt. 19:5). The translators drop a footnote which says: "Greek *one flesh" (not* "Some ancient authorities," or, "Some manuscripts say *one flesh")!*

That neutral word "one" permitting, the modern mind completes the sentence by saying "one person," or one in the personal realm. We are then off into the wild blue yonder, where there is an imagined convergence of the *telos* of medical technology with the direction of humankind toward God. Both go, it is believed, to the same omega-point. (Here Roman Catholics help themselves by a few citations from Teilhard de Chardin.) Along this path one shall never find good or sufficient reason to oppose any of the technical alterations of human parenthood that are rapidly becoming possible.

We need rather the biblical comprehension that man is as much the body of his soul as he is the soul of his body. The single word *sarx* in the "one-flesh" unity of marriage and parentage is sufficient to impel us to think-with the Jews and Christians in all ages who have affirmed a unity between the vocations of soul and body. They therefore affirmed the biological to be *assumed into* the personal and in some ultimate sense believed there is a linkage between the love-making and the life-giving "dimensions" of this one-flesh unity of ours.

AID (artificial insemination with donor) is only the first breach. Already at the University of Michigan, women have been impregnated from semen that had been frozen two and a half years before—to test whether this could be done without producing defective off-spring because of mutations or damage occurring during that period of time. Before considering the implications of this for human parenthood, one cannot pass over the fact that these procedures can be criticized as unethical human experimentation. Is it possible that these women were informed of the test to be made on them and on their children, and still consented to it? If not, the researchers were twice guilty—for violating the women's consent and for experimenting on an unborn child. If the women gave an understanding consent to be impregnated with aged frozen donor semen, they too were guilty of deliberately exposing the child they said they wanted to unknown hazards. Of course, the experiment had been carried out on animals, with no mishaps. But the step of passing from animal trials to human trials is a risky one, and it is spread upon the whole of medical ethics that this step ought to be taken only, or at least first, with adult volunteers. That, of course, is impossible in this case, except on the mother's part. If there is a child already conceived requiring genetic surgery, a parent can with fear and trembling consent to such investigational surgery in his behalf. However, a parent cannot legitimately submit a child who is as yet a hypothetical nothing to additional hazards for the sake of the accumulation of knowledge. Nor can he submit an existing child to purely experimental research.

Because we ought not to choose for a child—whose procreation we are contemplating—the injury he may bear, there is no way by which we can *morally* get to know whether many things now planned are technically feasible or not. We need not ask whether we should clone a man or not, or what use is to be made of frozen semen or ovum banks, or what sort of life we ought to create in hatcheries, etc., since we can *begin* to perfect these techniques in no other way than by subjecting another human being to risks to which he cannot consent as our coadventurer in promoting medical or scientific "progress." The putative volition of the child we are trying to learn how to manufacture must, anyway, be said to be *negative,* since researchers who work with human beings do not claim that they are ever allowed to ask volunteers to face possibly suicidal risks or to place themselves at risk of grave deformity.

By contrast with these steps in human research, and to Huxley's predestinators, it must be said that even Calvin's God was a more loving Father. Many of these proposals do not only remove parenthood as a basic form of humanity. They also would do violence to every spiritual aspect of human covenants and to the covenant with our Father in Heaven from whose name the standard for all fatherhood on earth is taken.

On the question of parenthood while men are on earthly pilgrimage, no Christian or Jew, it seems to me, can subscribe to proposals (whether "wise" or not in their consequences) that the human female, being a sophist-bird, should lay her eggs in concept and not in reality at all. The religious and moral traditions of the West cannot subscribe to the sweeping notion that thought is superior to the reality of the life of the flesh, a notion manifest in most of these dualistic "mind in dominion over body" proposals.

Concerning some or many if not all of these procedures for modifying the future of man, we have agreed that it can correctly be said: "When we take a comprehensive view of all this, it is clear that neither the biological nor the personal dimension is absent. Procreation has only been debiologized." But to be debiologized and recombined in various ways, parenthood must first be broken or removed. When the transmission of life has been debiologized, human parenthood as a created covenant of life is placed under massive assault and men and women will no longer be who they are. Mankind will no longer be, for man is no more nor less than *sarx* (flesh) plus the Spirit of God brooding over the waters.

This is no wedge-argument. It is not to say that once AID gains widespread social acceptance we will have influenced ourselves, individually and as a society, to go on to adopt more exquisite procedures for setting asunder *the person* with whom love is bodily communicated and *the person* with whom procreation of the one flesh of the child is accomplished. If the wedge-argument were at issue, I would certainly avail myself of it—since it is a good reason for anyone who has concern for *social ethics* or for *social practices.* The argument is rather that to assault the nature of human parentage by putting the bodily transmission of life completely asunder from bodily love-making (as in AID) is already essentially to have warranted many of these other procedures as well—perhaps the hatcheries, too. Such is already the implication of man's technically feasible debiologizing of himself and his parentage by donor insemination. Still it is also true that if, and only if, we highly respect the God-given nature of human parenthood are we apt to resist the powerful pressures in our technological civilization toward breaking apart and recombining, in a multitude of possible ways, the personal and the biological dimensions in human procreation.

The imagery and symbols we hear nowadays tell their own tale of what happens when human parenthood is completely *personalized:* biological "engineering"; the "manufacture" of chimeras; the "manufacture" of man; "hybridization" of human with other animal life to see what comes; the control of our biological "inventory"; "enforcing" birth control; sex for love, marriage for breeding; genetic "tailoring"; cures increasingly "wrought" upon or "done to" future generations as purely passive subjects. All logical successors to Kinsey's plumbing and electrical analogies: "inputs" and "outlets."

And, just think, it used to be fashionable to regard St. Augustine's imagery as unconscionably degrading of man, when he spoke of sexual intercourse as a man "plowing a field" and "sowing his seed"! One may sentimentally prefer pastoral imagery over that of an overweening technology. But what has to be said is that man has no homeland, humanism and morality no future, when man is reduced to either. Given a forced choice between being spiritually liquidated by one or the other environment, there is a good deal to be said in favor of horticultural as against engineering metaphors. Nineteen eighty-four is not an inappropriate date to choose for the biological apocalypse we are facing. Metaphorically speaking, the first opening in this direction occurred when we began to use a manufacturing term—"reproduction"—for procreation, human parenthood, and the transmission of the one-flesh of our lives through the generating generations of men.

No one should be happy about the merely bothersome function to which theology seems to have been reduced in the present age, nor with the wholesale abandonment of the full range of Christian ethical categories by its practitioners—an abandonment made in the name of "renewal," or so as not to stand in the way of widespread social acceptance of procedures that set entirely apart the person with whom we procreate and the person with whom we share the communications of an equally *incarnate* love.

Still, calculations of effectiveness do not determine the insights to which those who acknowledge our common moral tradition must bear witness. Even a bothersome witness may have some value in the in-between times—amid massive cultural breakdown. They also serve (it seems) who only stand and object to many of the overriding tendencies in the present age—upon good theological warrants. They also serve who only stand and ruminate upon the meaning and essential nature of human parenthood. They also serve who only stand and wait for the renewal of man. And they serve who wait for the emancipation of the transmission of fleshly life from techniques which came to serve and remained to master and to destroy, in the name of an unlimited "self-modifying" personal freedom. Perhaps God will let loose another St. Augustine upon this planet, and the wounds men have inflicted upon themselves in the modern period by "thing-i-fying" the *carnal* life may begin to be healed.

Questionable Aspirations to Godhood

The next point is theological. Men ought not to play God before they learn to be men, and after they have learned to be men they will not play God.

There are, of course, theologians who affirm that the judgment that men should not play God is only a "troglodyte and 'rear-view mirror' reaction," and not a principle that can at all clarify the meaning of responsible human decision. These theologians seek to elevate or assimilate any risk-filled, vital decision to playing a divine role.[28] Thus they avoid asking the critical questions about the meaning of man's *creaturely* responsibilities as a man or the real role of medicine and science as a human enterprise *serving* human life.[29]

We Americans are now familiar with the views of techno-theologians—so familiar, in fact, that many of us believe they actually are theologians and that in their writings they are using theological concepts or are doing religious ethics. They are rather to be deemed priests in an age in which the cultic praise of technology is about the only form of prophecy we know, or that can gain a hearing.

A recent article by the leading Roman Catholic theologian in Germany, Karl Rahner, probes profoundly the question of what, if anything, theology is going to be able to say about schemes for man's indefinite self-modification—so profoundly as to leave us at the heart of the problem.[30] Christianity—or for that matter, all the world's great religions—have taught, Rahner points out, that "man has always had the power to determine his permanent, everlasting orientation." But till now this power was "exercised almost exclusively in the area of the contemplative knowledge of metaphysics and faith," and in moral decisions opening man to the eternal. Now there has taken place an "historical breakthrough from theory to practice—from self-awareness to self-creation." This breakthrough by which men have now grasped and can fundamentally alter the roots of their own existence arises, Rahner believes, essentially from Christianity.[31] This judgment seems to be a spill-over from Rahner's esteem for man's powers of self-awareness and self-determination in contemplation. "Man is essentially a freedom-event. As established by God, and in his very nature, he is unfinished." From the fact that "freedom enables man to determine himself irrevocably, to be for all eternity what he himself has chosen to make himself," Rahner concludes that, in the practical order, the "creature" of this creative freedom is man himself. From the fact that "man determines himself in his innermost depths of his free actions," Rahner moves smoothly to an endorsement of man's modification of the "innermost depths" of his biological, genetic, psychological and historial existence.[32]

This sounds remarkably like a priestly blessing over

everything, doing duty for ethics. Rahner envisions "a hominized world, where man dwells as both the experimenter and the experiment—so that he can finally invent himself." "Man is experimentally manipulable," he writes, "and legitimately so." Since "evil is, in the final analysis, the absurdity of willing the impossible," Rahner concludes that "there is really nothing *possible* for man that he ought not to do" and that "there is no reason why man should not do whatever he is really able to do"—"what he ought not to do is—even today—impractical."[33]

This author seems to have an epiphenomenal and retrospective view of the task of theological ethics—at least in this article. "All we can do," he writes, "is discover the problems put to theology by human self-creation." He does not ask what questions theology puts to human self-creation. Something "becomes a theological question only after we have learned what self-creation can possibly accomplish and how."[34] If this is true, theological ethics itself is only a "rear-view mirror" reaction.

Running throughout is a sense of foreboding, a premonition of disaster. Rahner knows that there is such a pluralism of the human sciences that no single individual can grasp them all; and he does not believe that by means of teamwork and computers modern men can cope with this pluralism.[35] Yet he clings to the belief that men are wise enough to invent themselves. He knows that "these new laboratories have little in common with the older ones—the churches—where man was formed through the laborious amateur procedure of appealing to the free conscience of the individual." Yet he clings to the belief that men are good enough to form themselves from the bottom up. He knows that "purposeful, large-scale human self-creation may have irreversible, irreparable consequences in the future which future manipulation will be unable to undo." Yet he does not despair. Partly, this is because he appears to believe that automatic developments for good or ill would be "outside of the area of man's accountability before God." Partly, because he contradicts himself: it is, on the one hand, only "man's empirical self and not his spiritual, transcendental self, that is, his person as a subject of free activity" that will be affected by revolutionary, scientific self-invention; yet he knows these long-range plans "aim not at this or that aspect of man but at man as a whole," seeking "to penetrate the innermost depths of man in order to transform him."[36] Mainly, however, Rahner hopes on in faith because theologically, he believes that Christianity is "the religion of the absolute future."[37]

Therefore Rahner is sustained in the faith that, regardless of what may happen, "man's this-worldly self-creation has a positive relation to his self-opening to the absolute future." His theological-ethical imperative is: "make future self-creation a project worthy of man's absolute future—God himself." His confidence is that there are "laws" which "act as a control system to prevent [man] from straying irretrievably into the perverse and the absurd." (This accounts for Rahner's astonishing belief, mentioned earlier, that "there is really nothing possible for man that he ought not to do.") He asserts that "even the most radical human self-creation must be carried forward ontologically and ethically within a certain set framework which man has not fashioned and which can never be transcended." He calls for "possible guidelines for man's self-creation, which will always remain a venture into the unforeseeable."[38] But

Rahner gives no account of this set framework, nor does he adumbrate these laws or guidelines for man's self-creation, or attempt to say what sort of future self-direction of mankind would be worthy and what would be unworthy of man's absolute future.

The only "limits" man the self-creator comes up against are the facts that the world is "both created and fallen," both "judgment and blessing." Death and judgment manifest man's naked reality, and Rahner things it "odd to note how little mention is made of man's death and history's end in the usual utopian description of the future." The "law of death" sets a limit to what auto-creation can do and plan." But mention of these religious ultimates bracketing the whole human enterprise is productive of no direction for man's actions or abstentions in that enterprise. Instead, the reader must return to Rahner's rock of salvation: he simply "knows that man's history will always reach to God for its salvation or judgment, or else will cease to be the history of spirit-endowment persons."[39] We are not told what might be the parameters of a proper future history of spirit-endowed persons as they reach toward God, in contrast, say, to a sort of reaching toward God that could mean death.

This brings us to the heart of the problem—where we are left by Rahner, by Roman Catholic omega-pointers, and by Protestant theologians of secular, historical "hope" who collapse the distinction between being men before God and being God before we have learned to be men. Can any *articulate* meaning be given to the term "playing God" as a negative, critical norm of the moral life of mankind? Is this merely a pious notation or warning, having little or no determinate significance in deciding man's proper action?

Of course, from an understanding of the prohibition against "playing God" we can only expect to learn something about the *context* of decision and action that is properly human. The ethical principles and more specific moral rules that give shape to responsible action in the medical-technological context must be construed in detail and in their own terms. However, these particular decisions and actions begin to be shaped by virtue of the ultimate context in which the agent believes them to be located. Our concern now is to ask what general significance an ultimate religious context may have in determining appropriate human decisions and action. What does it mean, if indeed it means anything, to say that men should not play God before they have learned to be men and that when they learn to be men they will not play God?

Helmut Thielicke has written that the "borderline situation" is the place where men can learn the most about their lives.[40] AID is a "borderline" that, for example, is exceedingly propitious for learning or being reminded about the nature and meaning of human parenthood, the covenant among the generating generations of men. Similarly, taken as a whole, the proposals of the revolutionary biologists, the anatomy of their basic thought-forms, the ultimate context for acting on these proposals provides a propitious place for learning the meaning of "playing God"—in contrast to being men on earth.

At least this is the case if Donald Fleming, the Harvard historian, is at all correct in the account he gives of the outlook of the revolutionary biologists.[41] If his account is correct, never again can it be said—even by someone who does not accept the proposition—that there is no *meaning* to the regulative norm of not playing God. The central issue raised by the biological revolution is not the future of religion, as Fleming seems to suppose. The issue is rather the future of humanism—if any. Of man and his ethics—if any. Man becoming his own self-creator raises far more than vague religious trepidations.

A "full scale" revolution is in the making, not a minor one; it is directed at the foundations of mankind in the natural order. Fleming characterizes the engine propelling this revolutionary forecast by using such expressions as "a distinctive attitude toward the world," "a program for utterly transforming it," an "unshakable," nay even a "fanatical," confidence in a "world view," a "faith" no less than a "program" for the reconstruction of mankind. These expressions rather exactly describe a religious cult, if there ever was one—a cult of men-gods, however otherwise humble. These are not the findings, or the projections, of an exact science as such, but a religious view of where and how ultimate human significance is to be found. It is a proposal concerning mankind's final hope. One is reminded of the words of Martin Luther to the effect that we have either God or an idol and "whatever your heart trusts in and relies on, that is properly your God."

To speak more precisely, we have here a "messianic positivism"—of which there have been many versions in the modern period since that of Auguste Comte. Man seems to be, as Plato said, that most religious of animals who cannot avoid "faith-ing" even if he does so upon the firm foundation of unyielding despair over the biological deterioration of his species. To paraphrase Marxist apologists for utopia, whatever is real is bound to prove unreasonable in the end; therefore it is unreasonable now. And whatever is reasonable in the heads of the molecular biologists is bound to become real. There is nothing to do but to yield to this cult's Providence. Men are to erect a "good hope," even an "infinite hope," over the "corruption" of the existing world.

Not surprisingly, a whole new ethics follows from this surrogate theology. Human virtue and righteousness are now to be redefined in terms of the biological *summum bonum.* From this "faith-ed" onlook, "sloth" and "avarice" gain redefinition. That is not a complete list, but the beginning of a list of seven deadly sins! These vices now name those human attitudes that hinder chosen biological projects and render individual men worthless in the final biological assize. Anyone who does not love the absolute future of man's self-creation with all his heart and soul and mind and strength cannot be judged virtuous in any respect. No matter how humanly excellent he seems, his virtues are only "splendid vices." As Augustine said, if you want to tell whether a man is good or not, ask what he loves. The vices and virtues follow from this. "The new form of spiritual sloth will be not to want to be bodily perfect and genetically improved," say the revolutionary biological-*moralists* according to Fleming. "The new avarice will be to cherish our miserable hoard of genes and favor the children that resemble us."

The crucial point is, therefore, the future of humanism—if any. The issue we face in regard to the "messianic positivism" of some molecular biologists is not that they propose consciously to take up where *religion* left off. True, these biologists do not defend themselves against religion; they are not self-consciously anti-religious, as were some scientists decades ago. True, they now "subsist in a world where that has never been a felt pressure upon them." But it is not cultural or temporal distance alone which cause them to speak of religion in "the past tense." It is rather that these men now "sub-sist" in a world where formerly only God abounded. This reveals the anatomy of their thought, not where religion went.

Speaking of the man-God only in the future tense, these new messiahs (or, if one prefers, these new cultic bearers of an absolute biological future) must necessarily speak of humanism in the past tense as well. The crucial point is that they propose to take up where *humanism* left off. The point is not that, as Francis Crick put it, "there is going to be no agreement between Christians and any humanists who lack their particular prejudice about the sanctity of the individual."[42] The point is rather that there is going to be no agreement between this messianic positivism and any true humanism. A genuinely humanistic ethic holds to one version or another of that "prejudice." (Here I assume that "true believers" in the biological revolution are not out to prove that a humanistic morality must be rooted in some religious outlook.) The proposal Crick was discussing—to show that the facts of science are producing and must produce values that owe nothing to Christianity—was "the suggestion of making a child whose head is twice as big as normal." No "facts" and no facts of "science" produced that suggestion; men produced it—men who aspire to be gods over the whole creation. The point here is that a humanist or scientist whose practice of medicine or use of technical appliances is in the service of human life and from any remaining respect for the individual is also not apt to like that suggestion. It is not Christianity alone but man as well that the revolutionary biologists have left behind in their flights of grasping after godhead. No values at all are being produced, no values are even in conflict, if men come to subsist in a world in which the sole judgment to be made is—as Crick went on to say of that particular experiment—that men "simply want to try it scientifically."

And what can be made of Fleming's characterization of Joshua Lederberg as "at once maddened and obsessed by the nine-months phase in which the human organism has been exempted from *experimentation* and therapeutic intervention—such a waste of time before the scientists can get at us. But the embryo's turn is coming" (italics added)?

The most remarkable thing about the late H. J. Muller was not that his was an essentially Galtonian eugenics—the old eugenics newly garbed. The remarkable thing about this great scientist was that, although gloomy about mankind's biological prospects, he thought it worse to violate any human being and his free will for the sake of possible future biological improvements. He relied on man's voluntary and rational determination. He was an extraordinary humanist in his outlook; and it is an indication of the inhumane character of some of the procedures now contemplated that Muller's views can be called "subjective" and "imprecise." This is enough to show that man and his future may be at stake when scientific messianists can now calmly contemplate the utter removal or suppression of the human subject for the sake of their own version of some future state of affairs—a future state they propose to create *ex nihilo,* or out of biological components that do not take living men into respectful account. A number of the prospects now in our future and a number of the proposals now said to be mankind's overruling Providence would be a fundamental "violation of man" and of any conceivable form of humane ethics.[43]

In general, with regard to most of these proposals, the judgment of Gordon Rattray Taylor is much to be preferred to that of Donald Fleming or of the men whose views he reports. Taylor writes:

> I am therefore forced to the conclusion that society will have to control the pace of research, if it can, and will certainly have to regulate the release of these new powers. There will have to be a biological 'icebox' in which the new techniques can be placed until society is ready for them. This is not a conclusion to my taste at all. I do not feel in the least optimistic about our prospects of exerting such control without serious muddles and abuses. Nevertheless, the social consequences of what is in the pipeline could be so disastrous—nothing less than the break-up of civilization as we know it—that the attempt must be made.[44]

The final chapter of Taylor's book also details the non-Promethian views of some of the scientists whose work is preparing these options for us. That is, it details their own grave doubts of the value of doing many of the things they have made possible.

I would add only this word to Taylor's reference to a future time when we may have the moral and practical wisdom to take these techniques out of "deep-freeze" and use them rightly. We are not apt to come into the possession of such future moral wisdom if now we are utterly deprived of all grounds for judging that some things we *can* do *should* not be done—that we ought not to practice everything that *could* be technically accomplished. Taylor seems to allow for this in one sentence in his concluding paragraph: "knowledge, without the corrective of charity, hath some nature of venom or malignity."[45] That sentence bespeaks also the present, shaping influence of a *context* of decision and action in which men acknowledge that, for all their towering knowledge, they should not play God. We ought rather to live with charity amid the limits of a biological and historical existence which God created for the good and simple reason that, for all its corruption, it is now—and for the temporal future will be—the good realm in which man and his welfare are to be found and served.

The immanent providence of a morally blind biological technology decrees, of course, that men-gods *must* do what they *can* do. In the end, Fleming voices the justifiable sullenness of the men of the future who are to be worked over by our new deities and creators. "We the manufactured would be everybody and we the manufacturers a minority of scientists and technicians." But this will come, he predicts, with a gradual adjustment of our values "to signify that we approve of what we will actually be getting."

Against such a relentless predestination (and predetestation of man), living men need to count on something happening—arbitrarily, I suppose, or gratuitously. There may come an "e.g." that is out of place; this means an *"ex. gr."*— *"ex gratia"*—an act of *grace* which protrudes to remind us who we are as men. I suppose there will come a man who (like Dostoevski's undergroundling) will simply "stick out his tongue" at the whole scheme of things that the messianic positivists are preparing for his benefit. Despite all the spiritual pressure upon him to conform to this secularized theology, he will stick out his tongue at the conception that "spiritual sloth" means not to want to be bodily perfect and genetically improved by the priestly ministrations of biological science. He will revolt against this talk of becoming largely an "artificial man," or freezing himself to be resurrected, or causing a baby's brain cells to cleave again in order to double its size. Granting that he is the "distracted possessor" of a rather intolerable "identity," he will prefer his own to that of another, or to one created by pharmacology. A brain transplant will prove no more appealing than the transfer of his consciousness to a computer so that he can live forever (or until some mechanical breakdown). Rather than remain the purely passive patient of spectacular medical prowess, he may stick out his tongue at the idea of a fifth transplant, or the implantation of a fifth vital organ, and *freely and responsibly choose to die.*

I assume that man will refuse to live in a Crystal Palace or devote his every purpose to becoming a "card carrying cadaver." Man was not born from a chemist's retort. The scoundrel. To the chemist's retort he will not return. God or no God. Crick or no Crick.

Remove the messianic faith that has been added to scientific capabilities, however, and there remains ample room for significant interventions that are not only morally justifiable but morally required of men who have come to possess these capabilities. It has always been the teaching of our western religions that in *procreation* or the transmission of human life (block that manufacturing metaphor, "reproduction"!) men and women stand at the point where they should assume responsibility for future generations of mankind by assuming responsibility for their children. This would cut across a number of presently cherished social practices.

Doubtless the ethics of the future will not be the same as the ethics of the past. But the sine qua non of any morality at all, of any future for humanism, must be the premise that there may be a number of things that we *can* do that *ought* not to be done. Our common inquiry must be to fix upon those things that are worthy of man from among the multitude of things he is more and more capable of doing. Any other premise amounts to a total abdication of human moral reasoning and judgment and the total abasement of man before the relentless advancement of biological and medical technology. I do not believe that men should enslave themselves to an acknowledged minority of scientific saviors, or any man make himself willing to reduce another fellowman to a "thing in the world" over whom benefits are to be "wrought," while unfurling the banner of man's future triumph over natural forces. Calvin's God was never so offensive to the dignity of man.

This is the edification to be found in the thought that we should not play God before we have learned to be men, and as we learn to be men we will not want to play God.

A Penchant for Species-Suicide

Paradoxically, an ethical evaluation of the suppressed motive, the hidden or not-so-hidden tendency (and the end result) of a number of these schemes for the improving self-modification of man should be the same as the moral judgment we deem appropriate for acts of suicide. It is a special form of suicide we have been talking about: suicide of the species. We have followed only one of the lines of action moving in this direction —the control of the future of man through genetic manipulation and the alteration of human parenthood. There is another line of action which we have not traced out but which converges with the one we have followed. The end of both ways of making radical changes in humankind can only be described as the death of the species and its replacement by a species of life deemed more desirable. That I take to be similar to the inner motive and action of any suicide.

This second line of action concentrates not on genetic or reproductive modifications of the species but on the refabrication of the individual. Refabrication is to be accomplished by biology, of course, but also by spare-parts surgery so radical that the goal of a continually repaired man becomes an alluring vision. It is to be accomplished by spare-parts surgery, but also by pharmacology; by the prevention of aging, but also by making "cybogs" (combining biology with cybernetics); by the cure of all of man's ills even if that requires resurrection in the twenty-second century (cryonics), but also by transferring human consciousness to machines and by the perfection of computers that can teach and modify themselves in desirable ways if we cannot; by genetic and pharmacological control of a man's inner moods and powers, but also by remaking the present weak and distraught human self-awareness (the result, at least in part, of the environment's dominion over the individual) by establishing man's dominion over the environment.

All this adds up to man's limitless dominion over man. Who will be the Creator and who the creatures, who the masters who the slaves, who the miracle workers and who the things, at the end of these converging lines of development?—these are the questions we have already considered. The final moral perspective which, however subtle and paradoxical, we have now to examine arises from the fact that these proposals for man's assumption of control over his own evolution or (better said) over the next stage in *evolution* as such, display a strange penchant for species-suicide.

A fine example of this is the enthusiastic plea sent forth to every person now alive to join in the search for mankind's long-range goals—a plea made by Professor Gerald Feinberg, a physicist at Columbia University, in his book *The Prometheus Project*.[46] It is a good example of a contemporary intellectual's penchant for species-suicide precisely because Feinberg (like H. J. Muller among the geneticists) is a true humanist in his moral impulses and values.

Men now have the power to do things that influence the whole of mankind, not simply enclaves of the species. The effects of these actions can also foreclose a whole set of possibilities for an indefinite future. For this reason all men—or as many as possible—should take part in the setting of long-range goals having such encompassing consequences. The activity of participating in the determination of man's future for all time to come will itself be "spiritually ennobling." Prophecy again will become an honorable vocation, and one to which every man is called. "The goals of humanity" should "proceed from all of humanity."[47] Thus, the democratic humanism that undergirds Feinberg's plans to pool human insights is quite remarkable.

Moreover, he is a veritable Pascal come to judgment. The "emergence of human consciousness has finally enabled matter to become aware of itself." For the qualitatively unique "self-awareness"—that is, for man—to fail to "provide a plan and purpose for at least our part of the universe" would be to "fall short of our full humanity."[48] The whole material universe, which has no purpose, will not resist the purpose or purposes self-consciously adopted by men, since "nature neither knows nor cares, while man does, and therein lies his advantage."[49] The only thing that can stop us might be the sometimes unforeseen consequences of our own actions, but that difficulty can be overcome. "If only we agree on our goals, our technology can do the rest."[50]

A kind of existential humanism leads Feinberg to identify the sources of human misery that most need to be corrected, if we are to make all men glad and free. The flaw in "the human condition" lies at the heart of man's self-awareness of the human condition. "The most serious flaw" is our *finitude,* or rather the fact that "we are conscious beings *aware* of our own limitation." We are (1) distressed by the fact that we "lack the power to do the things we want" and (2) beset by "the specter of impending death, which always threatens to put an end to all our thinking and doing."[51] These are the two crucial alienations of self-aware finitude.

A partial solution to the first is to agree on our long-range goals and get to work implementing them, putting behind us the "long-range plans" of the great world religions of the past which counselled men to limit or suppress their desires. The solution to the second—a problem—arising from our consciousness that we are dying men—may call for a more radical reconstruction of man. If we ought not to try to keep our expectations finite, our accomplishments may still be limited—at least limited for any of the workers or any one of the generations of workers upon the City of Man. "I do not know whether the opportunity to exercise our abilities over an indefinite time period will itself be an answer to the unhappiness over our finitude," Feinberg remarks. "We will have to wait and see. If this should not prove the case, then some other kind of reconstruction appears called for to deal with finitude."[52]

It is evident from the beginning, however, that Feinberg is a strict reconstructionist. He asserts, it is true, that the Prometheus Project need not focus on "fixing the flaws in the human condition"; he affirms that there may be goals that are "an expression of positive features in man."[53] Still it is evident that the "Andreas fault" that causes all the earthquakes in a finite being conscious of his own finitude and death is the threat most in need of correction, or in need of an assuagement as yet unheard of in the entire history of humankind.

With becoming modesty and in the spirit of democratic humanism, Feinberg does not exclude from the Prometheus Project anyone who believes that long-range goals should be set "within the framework of what man is." His own opinion, however, is that "we should not wish to be bound by this restriction." To leap over the parameters of what man is is no incidental matter. To do this goes rather to the heart of the "proud and lonely distinction" of man's being in a universe where consciousness—or *our* consciousness—has dawned nowhere else. "My own feeling," Feinberg writes, "is that the despair of the conscious mind at the recognition of its own finitude is such that man cannot achieve an abiding contentment in his present form *or anything like it.* Therefore, I believe that a transformation of man into something very different from what he is now is called for. . . . [T]hose of us who find the human condition basically unsatisfactory must seek other methods as well."[54]

There are a number of things that are going to "come about" within the next century, technical advances that do not go very far beyond the known which men can seize and direct to their long-range goals. Feinberg seems particularly impressed with the possibilities of (1) the creation of intelligent and superintelligent machines, and their association with organic systems; (2) the con-

quest of aging; (3) the increase of intelligence by biological engineering and controlled environmental
changes, and the expansion of consciousness by a variety of means.

   1. There is no reason why a machine cannot be intelligent, even more intelligent than its creator—just as there
is no reason to deny that a man is intelligent because
"he is born of his mother and instructed by his teach-
er."[55] Superintelligent machines can provide alternative
sources for innovation in the creative activities hitherto
known as human, and machines can provide judgments
of merit as well.

They can be "functionally" intelligent in providing
and making choices. High-speed computers combined
with organic systems (cybogs) can enormously extend
the range of our thought and activity, whether we only
redesign "some aspects of the workings of bodies and
brains" of "already existing people, preserving that part
of their personality that makes them unique" or proceed to design entirely new individuals. Feinberg pon-
ders whether these things should be called "willful
machines," and says no, because we know so little about
the will. Still he believes that "the existence of intelligent machines would do something to relieve the spiritual loneliness that comes from being the only sentient
beings in the universe."[56]

   2. The process of aging will be defeated. Man may live
a hundred or a thousand years in good condition. Men
need not waste by early death the experience gained in a
lifetime. The length of life would help alleviate the fear
of death, although, as Feinberg suggests, fear of death
from aging may only be replaced by fear of death by
accident. And, under the conditions of the future, death
may be a thing more to be feared. Population growth

could still be brought to zero if women had no more than two children and if everyone died sometime. We also need to think about the psychological effects of spending only a disappearing "moment" of time out of centuries of lifetime in raising children. To counteract the bad effects of people getting too "set in their ways," we could "produce people who did not attain unalterable views through experience," or we could rearrange society so that "intellectual activity was the work of the relatively young, and the elderly were left to enjoy physical pleasures."[57] Have not philosophers long ago told us that youth is too wonderful to be wasted on the young?

3. The increase of intelligence is an obvious goal. As long as this means an effort to overcome the shortcomings of human heredity "to meet the prevailing human norms, there is not much question of choice involved." It would be one of Feinberg's "developmental goals," which "involve the intensification or fulfillment of something already to some extent present in humanity." Mankind needs to press on to "transcendent goals" which "require the creation or achievement of something qualitatively new." Advances now on the horizon are only suggestive of these possibilities—advances such as the increase of intelligence by biological means and the expansion of consciousness by environmental, electrical, and chemical stimulation. There can be "a detailed ethical alternative to ordinary life involving the systematic use of drugs."[58]

Possible long-range plans for the human race include a "kind of Faustian search for a yet undiscovered state of bliss." Yet there is the danger that mankind could get caught in a cul-de-sac. For example, it is possible simply to stimulate the pleasure centers or to induce mental

states artificially—the result corresponding to any imagi-
nable human experience without the individual undergoing the experience itself or engaging in any activity at
all.[59] (That might only be a diversion introduced into
the copulation explosion.[60]) For these reasons, it is of
paramount importance that we think about our long-
range goals, and proceed cautiously in order, if possible,
to allow ourselves to reverse course. (Men could get lost
in a drug culture or in mystical experience that is illu-
sory.)

Feinberg's chief proposal for eternal, universal life in
the end-time of humanity is an extension of consciousness which these initial achievements only adumbrate
and may help to make possible. By adding to these lines
of action we could open the way to this final goal. Then
alone will the faults and despair at the heart of the
human condition be overcome. Perhaps we will one day
create some form of artificial intelligence by which
natural phenomena can be directly and immediately
known. (This would be a power similar to the one at-
tributed in former times to the angels, different only in
the object of knowledge believed to be worthy of intellectual attention.) So that there may be "no more feeling of separateness than now exists in the mind of one
person," we must achieve the fusing of two minds into
one. "The first attempts at merged consciousness would
probably be temporary; after a time the individual
minds would separate to work independently again. This
would give them a chance to see whether they preferred
individuality or the merged state. Eventually, a group of
minds might choose to remain permanently united, thus
creating a new being."[61]

Ultimately we might want to go beyond the human
race to create a "commonwealth of conscious beings"—

"be they humans, Martians, dolphins or IBM 137000."
We can count on it that the goals of beings "that have
only consciousness in common would also relate to consciousness."[62]

This is the urgently desired cure for man's finitude,
for man's distressing "inability to accomplish all that he
wills." By extending consciousness through merger with
all conscious being, we shall "internalize more of the
outside world," "bring more of the totality of experience into the direct control of our consciousness," remove the source of human unhappiness by attacking the
conflict between the mind and the world. The intransi-
gence of the external world, the limitation of consciousness to a small part of the world of matter and to a
limited time, and—most important of all—the fact that
there is now "not one consciousness but many" separate
from one another: these sources of human grief and
grievance will finally be abolished. "A poet might say,"
Feinberg concludes, "that consciousness must be co-
extensive with creation. . . . no more finite than the
universe itself." What then would consciousness be conscious of? It would be *conscious of consciousness*—of
"aspects of itself." When consciousness becomes a universal attribute, then "it can at last play the role to
which it is entitled by the value we put upon it."[63]

Is there a God? "There is NOW," came the answer.

At the same time, it is evident that, in the end and far,
far earlier in the converging lines of action leading to
man's radical self-modification and control of his evolutionary future, many of these proposals must simply be
described as a project for the suicide of the species. The
momentum behind them is despair over man as he is.
Because those who come after us may not be like us, or
because those like us may not come after us, or because

after a time there may be none to come after us, mankind must now set to work to ensure that those who come after us will be more and more unlike us. Because the earth will finally become uninhabitable, its present inhabitants must "go to meet the planets," not simply by space travel but through radical genetic alteration to ensure that we shall be the forebears and progenitors of the strange and unearthly forms of life that may be able to exist on the planets—even though the only relation between us and "them" will be that we have been their progenitors. Our hope is fixed in nothing less than these futures. In this there is at work the modern intellect's penchant for species-suicide.

In the first genesis, men with expectation high savored knowledge and God-head, death following. In the second genesis men with expectation high savor death to the species, of man as he is, God-head following.

Our ethical judgment—not upon science but upon these science-based proposals put forward in the context of a propensity for the destruction of the parameters of human existence—must be the same as the judgment we would render upon suicide on a grand scale: suicide of the species, in the expectation of godhood following (or some reasonable facsimile thereof).

The Space Child with Brow Serene, when she appears, will discern the flaw in all the forms of modern dualism and in every scheme for usurping dominion over man and his future. With Brow Serene the Space Child will declare, as Augustine said of the Manichees, these men do not accept with good and simple faith that for one good and simple reason God created the world—because it is good.

Notes

Chapter 1

This chapter first appeared in John D. Roslansky, ed., *Genetics and the Future of Man* (Amsterdam: North-Holland Publishing Company, 1965).

1. New Brunswick, N.J.: Rutgers University Press, 1963.

2. Comments on Genetic Evolution, in Hudson Hoagland and Ralph W. Burhoe, eds., *Evolution and Man's Progress* (New York: Columbia University Press, 1962), p. 41. "The troubled history of Utopian education warns us to take care in rebuilding human personality on infirm philosophy" (Joshua Lederberg, "Biological Future of Man," in *Man and His Future*, a Ciba Foundation Volume [London: J. and A. Churchill, 1963], p. 270). "It must be pointed out rather emphatically that the genetic consequences of a eugenic program based on faulty or inadequate genetic knowledge could, in themselves, be as dangerous to our genetic endowment as radiation. It seems crystal clear that the implementation of some of the more bizarre eugenic recommendations of several decades ago would have been the worst sort of folly" (Bruce Wallace and Theodosius Dobzhansky, *Radiation, Genes, and Man* [New York: Henry Holt, 1959], p. 191).

3. Donald M. MacKay, in the discussion of eugenics and genetics in *Man and His Future*.

4. Frederick Osborn, "The Protection and Improvement of Man's Genetic Inheritance," in Stuart Mudd, ed., *The Population Crisis and the Use of World Resources* (The Hague: Dr. W. Junk Publishers, 1964), p. 308.

5. Theodosius Dobzhansky, *Mankind Evolving* (New Haven: Yale University Press, 1962), p. 332.

6. Ibid.

7. Lederberg, "Biological Future of Man," p. 264.

8. James F. Crow, "Mechanisms and Trends in Human Evolution," in *Evolution and Man's Progress,* pp. 9, 11.

9. See, for example, H. J. Muller, "Our Load of Mutations" in *The American Journal of Human Genetics* 2 (June 1950): 165.

10. Dobzhansky, *Mankind Evolving,* pp. 301, 330.

11. Hampton L. Carson, *Heredity and Human Life* (New York: Columbia University Press, 1963), p. 137. "Any genetically determined trait, no matter what, which makes a man a better producer of a large and healthy family favors this particular line of descent just because this line makes a relatively large contribution to the composition of the gene pool of the next generation. Actually, the process can go on, and usually does, without any active struggle between the parties concerned."

12. H. J. Muller, "Should We Weaken or Strengthen our Genetic Heritage?" in *Evolution and Man's Progress,* p. 23.

13. In the discussion printed in *Evolution and Man's Progress,* p. 57.

14. H. J. Muller, "The Guidance of Human Evolution," in *Perspectives in Biology and Medicine* (Chicago: University of Chicago Press, 1959) 3 (Autumn 1959): 13.

15. H. J. Muller, *Man's Future Birthright* (University of New Hampshire, Feb., 1958), p. 14.

16. Ibid., pp. 128, 133, 143, 144.

17. Muller, "Our Load of Mutations," p. 169.

18. Ibid., p. 150.

19. These terms are used by H. J. Muller, "Means and Aims in Human Genetic Betterment," in Tracy M. Sonneborn, ed., *The Control of Human Heredity and Evolution* (New York: The Macmillan Co., 1965). "Nano" designates a scale a thousand times smaller than "micro."

20. Wallace and Dobzhansky, *Radiation*, p. 32.

21. Augustine, *The City of God*, bk. 12, chap. 13. *The New York Times*, Dec. 26, 1964, reported an incident that happened on Christmas day in Louisville, Kentucky, which may be compared with the realization of genetic improbabilities. In a game of bridge it happened that *all four players* were dealt hands each containing the 13 cards of the same suit. According to the World Almanac, the chance that *one player* in a bridge game will get 13 cards of the same suit is one in 158,753,389,900. The odds against all four players getting the same suit are not given. These odds must be tremendous, yet it happened. When one remembers that such highly unlikely events can happen early in a series of deals rather than after an indefinite number of deals, then it would seem entirely possible that individuals other than "identical twins" have already been "dealt" the same genotype in the course of human history; and that these genotypes may be called upon to reappear again.

22. The ancients were, quite correctly, possessed by a "spirit of melancholy" (Nietzsche) over the eternal recurrence, because it entailed the inexorable disappearance of the best, as well as its return in the cycles. Having derived from the Christian doctrine of providence the idea that human history has *linear* significance, if it has any at all, a modern geneticist is likely, instead, to be afflicted with unmitigated gloom as he faces the prospect of irreversible linear genetic degeneration.

23. See Crick in the discussion recorded in *Man and His Future*, pp. 275-76.

24. It is quite obvious that the voluntariness upon which Muller's program is based has more substantial foundations in his estimate of man than the basis he once asserted it to have in the "genetic feedback" of servile traits a compulsory program would produce. "A dictatorship," he wrote, "though it might hoodwink, cajole and compel its subjects into participation in its [eugenic] programme, would try to create a servile population uncomplainingly conforming to their ruler's whims. *That would constitute an evolutionary emergency* much more immediate and ominous than any gradual degeneration occasioned by a negative cultural feedback." ("Genetic Progress by Voluntarily Conducted Germinal Choice," in *Man and His Future*, p. 257; italics added.) One has to be a scientific *humanist* even to know what constitutes an evolutionary emergency.

25. *The World View of Moderns*, University of Illinois 50th Anniversary Lecture Series (Urbana, Ill.: University of Illinois Press, 1958), p. 15.

26. "Better Genes for Tomorrow," in Mudd, *Population Crisis*, p. 323.

27. Wallace and Dobzhansky, *Radiation*, p. 194.

28. Cf. Pascal's "thinking reed" passage in *Pensées*, 347. Muller writes: "By working in functional alliance with our genes, we may attain to modes of thought and living that today would seem inconceivably god-like. In this expression the word 'thought' had advisedly been set before 'living.' For

thought is the distinctive and central mode of existence of man, the new mode of expression of the genes, and in the beings who succeed us if we win out, thought will ever more truly come into its own" ("Man's Place in the Living Universe," an address at Indiana University, June 9, 1956 [Indiana University Publications, 1956], p. 24).

29. First published in 1956; Torchbook edition (New York: Harper, 1959), p. 13.

30. Ibid., pp. 61, 72-73.

31. Ibid., p. 74. In the "body" or "fellowship" of scientists "the power of virtue" must be operative: "All scholars in their work are . . . oddly virtuous. They do not make wild claims, they do not cheat, they do not try to persuade at any cost, they appeal neither to prejudice nor to authority, they are often frank about their ignorance, their disputes are fairly decorous, they do not confuse what is being argued with race, politics, sex or age, they listen patiently to the young and to the old who both know everything" (p. 75).

32. Ibid., in order, pp. 77, 80, 81, 83.

33. Ibid., in order, pp. 13 (italics added), 81.

34. It is manifestly absurd for Bronowski to explain William Wilberforce's successful opposition to the slave trade by saying: "He had at bottom only one ground: that dark men are men. *A century and more of scientific habit by then had made his fellows find that true"* (Ibid., p. 60, italics added). After all, Wilberforce was a Bishop! Similarly, can anyone believe the following passages to be at all adequate accounts of two great moments in the history of British and American liberties? (1) ". . . not the high talk about the divine right of kings, and not the Bill of Rights, but their test in experience. England would have been willing to live by either concept, as it has been willing to live by Newton or by Einstein: it chose the one which made society work of itself, without constraint" (p. 55). Surely, the extension, through centuries of struggle, of the meaning of the Magna Carta's "No freeman shall . . ." was a process by which men apprehended more clearly and put into exercise the meaning of liberty which was 'otherwise known.' (2) "We see it [the test of experienced fact] cogently in the Declaration of Independence, which begins in the round Euclidean manner: 'We hold these truths to be self-evident,' but which takes the justification for its action at last from 'a long train of abuses and usurpations': the colonial system had failed to make a workable society" (p. 56). Surely, any school child knows whence our founding fathers took the justification for their actions at last, and that, without knowledge of the inalienable rights of man bestowed by nature's God, they could not have known abuses to be abuses, or usurpations to be usurpations.

35. "Guidance," p. 11.

36. "Our Load of Mutations," pp. 146, 171.

37. Ibid., p. 146. Cf. also Muller, "Should We Strengthen or Weaken our Genetic Heritage?", p. 27. It does not seem a sufficient answer to all this to reply: "Norway rats . . . have been kept in laboratories since some time before 1840 and 1850. . . . But it does not follow that laboratory rats are decadent and unfit; nor does it follow that the 'welfare state' is making man decadent and unfit—to live in a welfare state!" (Dobzhansky, *Mankind Evolving,* p. 326).

38. Muller, "Better Genes for Tomorrow," p. 315.

39. Hannah Arendt, quoted in a *Worldview* editorial, Sept. 1958, p. 1.

40. In an article entitled "Sex and People: A Critical Review" (*Religion and Life* 30 [Winter 1960-61]: 53-70), I sought to apply the edification found in Christian eschatology in refutation of certain genial viewpoints sometimes propounded by Christians on the basis of a doctrine of creation. These Christians hold that religious people *must* believe that God intends an abundant *earthly* life for every baby born, and that we would deny His providence if we doubt that world population control, combined with economic growthmanship, can finally succeed in fulfilling God's direction of human life to this end. Such a belief is secular progressivism with religious overtones. Taken seriously enough, it can lead, as easily as any other utopianism can, to the adoption of any means to that end, the control of the world's population. In essence, an independent morality of means, or righteousness in conduct, is collapsed into utilitarianism when the *eschaton* or man's supernatural end is replaced by any future *telos*.

41. The language of this paragraph reflects that of H. Richard Niebuhr, *The Responsible Self* (New York: Harper and Row, 1963).

42. "A Free Man's Worship." There is less posturing in Muller's despair, more in the optimism that floats over this despair, than in Russell.

43. "Ethics in International Relations Today," an address delivered at Amherst College, Dec. 9, 1964; quoted from *The New York Times,* Dec. 10, 1964.

44. "A Christian Approach to the Question of Sexual Relations Outside Marriage," in *The Journal of Religion* 45 (Apr. 1965): 100-18.

45. Gerald Kelley and John C. Ford, "Periodic Continence," in *Theological Studies* 23 (Dec. 1962): 590-624.

46. In these and other statements in the explanation in the text above, I may seem no longer to be within hailing distance of normative Judaism, or of what follows from intending the world as a Jew. There is a profound, *formal* analogy, however, to be taken into account if anyone wants to understand the basic theological ethics that is, or should be, controlling in the specific teachings of Judaism. Normative Judaism, or at least the theology of the Jewish Scriptures, also understands creation from the point of view of convenant. From the center of those events in which God created a negligible tribe to be His people, they understand his will in creating anything else out of "nothing," convenanting with the sun and moon and stars. And from the point of view of God's faithfulness they interpret the fidelity or steadfastness manifest anywhere in the world, while depending on God's Messiah more than on anything else to prove this steadfastness.

47. "Means and Aims," p. 117.

48. "Guidance," p. 26.

49. "One must face the fact that there is eventually bound to be a conflict of values," said Crick in the discussion in *Man and His Future,* p. 380. "It is hopeful that at the moment we can get a measure of agreement, but I think that in time the facts of science are going to make us become less Christian." The subject of this discerning remark was the disagreement between Christians, with "their particular prejudice about the sanctity of the individual," and those who "simply want to try it scientifically"

(whom, strangely, Crick called "humanists"!). But when it came to any of the finer points (such as those discussed in the text above) which anyone who presumes at all to take up the subject of Christian ethics makes himself responsible for knowing, Crick could only use a very blunt and unanalyzed notion of people's "right to have children" which, he asserted, is "taken for granted because it is a part of Christian ethics." Against this supposed notion, he wanted to "get across to people the idea that their children are not entirely their own business and that it is not a private matter" (p. 275). And in the discussion in *Evolution and Man's Progress* Professor D. H. Fleming, historian at Harvard, expressed some degree of reluctance about having science assume the moral leadership of mankind by the adoption of Muller's proposals, because, he said, this would "represent a passing over to science of the traditional role of religion as the fountainhead of restraints upon pleasurable conduct" (p. 65). To which the appropriate reply is: "Goshallhemlock!"

50. A great deal has been said about the weak reasoning—psychological, sociological, etc.—which Pope Paul VI used to support the traditional Roman Catholic prohibition of artificial conception control in his encyclical *Humanae Vitae* ("On the Regulation of Birth"), July 25, 1968. The crucial point, however, was the Pontiff's mistaken conception of the viewpoint he wished to reject. "Could it not be admitted," he asked rhetorically, "that the finality of procreation pertains to the ensemble of conjugal life, rather than to its single acts?" (par. 3). That one word *ensemble* betrays the fact that for the Pope the only alternative to holding the procreative and the communicative goods or "finalities" together in every single act was to hold them together in a whole series of single acts—in ensemble, first one and then the other, in a couple's conjugal life. That is still too much bound by single-act analysis. It is quite another thing to say that the nurturing of love and the transmission of life (even when the latter is limited) are to be held together *within the covenant of marriage,* with *the person* with whom one has made marriage-covenant.

51. H. J. Muller, who favors phenotypic selection, describes the enormous difficulties in the way of perfecting methods of genotypic change in "Means and Aims." In the advancement of science toward direction or change of the germ cells themselves, Muller believes "there may be in time a race between genetic surgery and robotics, and we may find that 'this old house will do no longer' " (p. 109). I take him to mean that a new type of man may be as easily made as present man can be remade by direct action on his genes. Neither, for Muller, is "utterly visionary." Since both robotics and the direction of mutation are visionary, however, Muller wants to proceed with parental selection by all the voluntary means presently available.

52. Michael Lerner, Professor of Genetics at the University of California (Berkeley), pointed out that animal breeders made little improvement until they "had clearly defined objectives," "used exceedingly high levels of inbreeding—the basis of breed fixation," and used "very expensive techniques in terms of genetic extinction, that is, in terms of preventing the reproduction of huge numbers of individuals in order to improve the trait of one or two percent" (in the discussion in *Evolution and Man's Progress,* p. 55).

53. Lederberg, "Biological Future of Man," p. 265 (italics added).

54. Muller, "Guidance," p. 37 (italics added).

55. "As in most defensive operations, it is dreary, frustrating business to have to run as fast as one can merely to stay in the same place. Nature did better for us. Why can we not do better for ourselves?" ("Guidance," p. 17). Thus, only progressive eugenics would be the equivalent of natural selection, which was phenotypic and preserved the genes of the strongest types.

56. "Guidance," p. 35.

57. Dobzhansky, p. 328; and Klein's comment in the discussion in *Man and His Future,* p. 280.

58. J. Paul Scott in the discussion in *Evolution and Man's Progress,* p. 48.

59. R. S. Morison in ibid., p. 64.

60. Donald M. MacKay in the discussion in *Man and His Future,* p. 298.

61. John F. Brock in ibid., p. 287. Or that, in view of the incredible diversity of opinions expressed by the scientists, it is impossible to know what we should try to educate people to do in making genetic choices (Medawar in *Man and His Future,* p. 382).

62. There is an exceedingly profound and open-minded discussion of artificial insemination, from the point of view of a Lutheran ethics, in Helmut Thielicke's *The Ethics of Sex,* trans. John W. Doberstein (New York: Harper and Row, 1964), pp. 248-68.

63. The Roman Catholic legal authority, Norman St. John-Stevas (*Life, Death and the Law* [Bloomington, Ind.: Indiana University Press, 1961], pp. 116-59), gives a good account of the theological, moral and legal aspects of this question. He leans in the opposite direction from the position suggested in the text above.

64. "Better Genes for Tomorrow," p. 336.

65. "Genetic Progress," p. 260.

66. *The World View of Moderns,* p. 26. Without some consensus on the ultimate question of values, he points out elsewhere, all man's cultural activities, no less than his germinal choices, would be at cross-purposes ("Guidance," p. 19).

67. *Radiation,* p. 330.

68. "Couples desiring to have in their own families one or more children who are especially likely to embody their own ideals of worth will be afforded a wide range of choice. They will be assisted by records of the lives and characteristics of the donors and of their relatives, and by counsel from diverse specialists, but the final choice will be their own and their participation will be entirely voluntary" ("Means and Aims," p. 122).

69. Ibid., p. 118.

70. Donald M. MacKay in the discussion in *Man and His Future,* p. 286.

71. *Heredity and Human Life,* p. 189.

72. Ibid., p. 188.

73. See Crow, "Mechanisms and Trends," p. 18.

74. Osborn, "Protection and Improvement," pp. 308-09.

75. See Wallace and Dobzhansky, *Radiation,* pp. 184-85. Since these authors had just cogently stated (perhaps without knowing it) the "rule of

double effect," I frankly do not understand their meaning in the following paragraph: "The importance one places on genetic damage depends, really, on the value one places on human life. If the importance of human life is absolute, if human life is infinitely precious, then the exact number of additional victims of genetic damage is not crucial. One death is as inadmissible as 100, 1,000, or 1,000,000. Infinity multiplied by any finite number is still infinity. Whoever claims that the number of genetic deaths is an important consideration in this problem claims that human life is of limited value" (p. 188). To the contrary, it is precisely because each human life has such value that it becomes important to take the numbers into account as one element in the proportion in situations where *not all can be saved.* Prudence is a matter of estimating the cost-benefit where infinite values (the lives of persons) are in conflict, where, e.g., persons in the present generation must be saved at the expense of persons in a future generation, or vice versa; and there is *no other alternative.*

76. *Man's Future Birthright,* p. 18. See also Muller's "Guidance," p. 8.

Chapter 2

An address given at a conference on "Ethics in Medicine and Technology" sponsored by the Institute of Religion at the Texas Medical Center and by Rice University (Houston, Texas; March 25-28, 1968). First published in Kenneth Vaux, ed., *Who Shall Live? Medicine, Technology, Ethics* (Philadelphia, Pa.: Fortress Press, 1970), p. 78-113.

1. "Experimental Genetics and Human Evolution," in *The American Naturalist* 100 (Sept.-Oct., 1966): 519-31, revised slightly and reprinted in the *Bulletin of the Atomic Scientists* (Oct. 1966): 4-11. Also see Lederberg's column, "Unpredictable Variety Still Rules Human Reproduction," in *The Washington Post,* Sept. 30, 1967.

2. This and subsequent quotations, unless otherwise noted, are from the version of Lederberg's article "Experimental Genetics," reprinted in the *Bulletin of the Atomic Scientists.*

3. Quoted from "Unpredictable Variety."

4. This remarkable experiment is described in Robert Briggs and Thomas J. King, "Transplantation of Living Nuclei from Blastula cells into Enucleated Frogs' Eggs," in *Proceeedings of the National Academy of Sciences* 38 (May 1952): 455-63.

5. See Thomas J. King and Robert Briggs, "Serial Transplantation of Embryonic Nuclei," in *Cold Spring Harbor Symposia on Quantitative Biology* 21, (1956): 271-90.

6. Quoted from *The Evening Bulletin* (Philadelphia), Feb. 23, 1968.

7. Thomas J. King, "Nuclear Transplantation in Amphibia," in D. M. Prescott, ed., *Methods in Cell Biology,* 2 (New York: Academic Press, 1966), pp. 1-36.

8. Marie A. Di Berardino and Thomas J. King, "Development and Cellular Differentiation of Neural Nuclear-Transplants of Known Karyotype," in *Developmental Biology* 15 (Feb. 1967): 123.

9. "The dormant storage of human germ plasm as sperm will be replaced by the freezing of somatic tissues to save potential donor nuclei."

10. Lederberg restates this objection: "[The] introverted and poten-

tially narrow-minded advantage of a clonish group may be the chief threat to a pluralistically dedicated species." In the same vein, Dr. Kurt Hirschhorn, President of the American Society of Human Genetics, declared at a meeting of the American Pediatric Society in Atlantic City, May 1, 1969, that the proposal to produce identical copies of certain special individuals "suffers from the same hazard" as an earlier, once-popular eugenic suggestion—to improve the human species by artificially inseminating hundreds of women with the sperm of "top notch" men. The hazard, he said, is that "both approaches are a form of inbreeding that would decrease the genetic variability of the human species" (Quoted from *The New York Times,* May 2, 1969).

11. "Particular interest attaches to the reproductive capacity of nuclear transplant frogs," writes Dr. Gurdon, "since this provides the most critical test of whether the nuclei from which they are derived are equivalent to the germ-cell nuclei in the range of differentiation that they can promote" (J. B. Gurdon, "Adult Frogs Derived from the Nuclei of Single Somatic Cells," in *Developmental Biology* 4 (1962): 256, 261-62, 268-70. His findings in the case of the infertile ones of his cloned frogs "are consistent with" the hypothesis that this was due to nuclear-cytoplasmic incompatibility and changes induced during the development of the donor embryo. That is, infertility may have been due to the fact that chromosome replication in the tissue-nucleus was out of phase with the cleavage of the cytoplasm. Dr. Gurdon counted as abnormal "only those frogs which fail to prove fertile and to give normal offspring [up to the average of the frogs raised in the laboratory] on at least three occasions." This indicates an abnormality in even the fertile ones: lower and more erratic fertility. Another abnormality (other than those leading to total infertility) was the fact that when frogs were cloned by nuclear transplantation from gut tissue (instead of *earlier* embryonic tissue) most of the *male-*producing donors gave rise only to sterile frogs. "This departure from equality in the number of female- and of male-producing donors," Dr. Gurdon comments, "suggests the possibility that endoderm nuclei from *advanced donor stages* are unable to give rise to normal male differentiation, so that sex reversal occurs." This is enough to indicate again the formidable technical difficulties in replicating a normal individual of any species. It is also enough to show the formidable moral question concerning the mishaps that will result from ever attempting to do this experiment on man. (See also J. B. Gurdon, T. R. Elsdale and M. Fischberg: "Sexually Mature Individuals of *Xenopus laevis* from the transplantation of Single Somatic Nuclei," in *Nature* 182: 64-65.)

12. The normal embryos obtained by King-Briggs could be explained in two different ways, as King himself pointed out: "Normal embryos obtained from transplanted nuclei can be explained equally well by assuming that a limited number of nuclei in a given cell population are totipotent, or that some nuclear changes that do occur can be reversed as a consequence of replicating in the cytoplasm of test eggs" (Di Berardino and King, "Development and Cellular Differentiation," p. 125).

13. See J. B. Gurdon, "The developmental Capacity of Nuclei Taken from Intestinal Epithelium Cells of Feeding Tadpoles," in *Journal of Embryology and Experimental Morphology* 10 (Dec. 1962): 622-40.

14. In the light of this remark, it would seem that the testimony of Dr. Lederberg and his Stanford colleague, Dr. Arthur Kornberg, before the Senate Subcommittee on Government Research on the resolution of Senator Walter E. Mundale (D-Minn.) to set up a National Commission on Health, Science and Society was less than forthright and more than disingenuous. "How the human species can foresee and plan its own future is the real subject of these hearings," said Dr. Lederberg. "No subject is more important." But the issues were too important to be left with a national commission. One may indeed doubt whether a commission "charged with making substantive prescriptions, after one year's study, about the biological policy of the human species," is the answer (quoted from *The New York Times,* Mar. 9, 1968). But who then? Both Drs. Lederberg and Kornberg abdicated giving any answer to this question. They seemed to deny that recent progress in genetics poses special moral, social or political problems for mankind. Both called for more research and more funds for research—for a public well-educated in the concepts of science and a scientific community richly supplied with data. In short, they called for a purely scientific answer to the moral questions arising from genetic and nucleic acid research, even if they did not quite claim that only scientists themselves should fashion public policy concerning these matters.

15. "Experimental Genetics," p. 530 n. 2.

16. I have quoted from "A Geneticist Looks at Contraception and Abortion," *The Changing Mores of Biomedical Research: A Colloquium on Ethical Dilemmas from Medical Advances,* held at the 48th Annual Session of the American College of Physicians, San Francisco, April 12, 1967, in *Annals of Internal Medicine* 67 (Sept. 1967): 25-27.

17. Cf. above: ". . . communication with the species, which is the foundation on which the unique glory of man is built."

18. Quoted in *Annals of Internal Medicine* 67 (Sept. 1967): 54-55.

19. *The Ethics of Sex,* p. 199.

20. Paul L. Lehmann, *Ethics in a Christian Context* (New York: Harper and Row, 1963), p. 99.

21. "Unpredictable Variety Still Rules Human Reproduction"; the *title* of this column is also of considerable significance.

22. Ei Matsunaga, "Possible Genetic Consequences of Family Planning," a paper read before a seminar of the International Planned Parenthood Federation (Tokyo, May 26, 1966), National Institute of Genetics, Mishima, Yata, Japan.

23. There is also mounting evidence from studies of human reproduction, including evidence gathered by rhythm clinics, that among pregnancies resulting from intercourse a number of days before ovulation there are an abnormal number of males. (See R. Guerrero, *Time of Insemination in the Menstrual Cycle and Its Effect on the Sex Ratio,* thesis, Harvard School of Public Health [Boston, Mass., January, 1968].) Thus, acts of sexual intercourse qua procreative are a series of "time acts," and to disturb this by a selection favoring (if only by "accidents") those days more remote from ovulation may, as a consequence, disrupt the proportion of males and females in the future population. If anything, the optimum good of the conceptus, the good in the nature of human parenthood, and the male-female balance would require selection in favor of intercourse at the time

of ovulation. This would require the use of temperature charts, etc., for *this* purpose, whenever procreation is wanted, and the use of assured methods of contraception during the remainder of the cycle.

24. Dr. Kass's letter was printed in *The Washington Post* on Nov. 3, 1967.

25. Dr. Lederberg's response appeared in *The Washington Post* on Nov. 4, 1967.

26. "Dr. Barnard Rejects Curbs on Doctors," by-line Victor Cohn, *The Washington Post,* Mar. 9, 1968.

Chapter 3

1. Leo Straus, "Jerusalem and Athens: Some Introductory Reflections," *Commentary* (June 1967): 45-57.

2. "What are we going to say of a gorilla whose brain has acquired a speech center through the admixture of human genes? Or, rather, what will we say *to* such a creature when it applies to enter the university?" Men might also mix animals and plants, producing a creature with a brain large enough to "indulge in philosophy" but with a photosynthetic area on its back so that it would not have to eat. (Rodney Gorney, "The New Biology and the Future of Man," *U.C.L.A. Law Review* 15 [Feb. 1968]: 304.)

3. This figure was probably translated to the Fahrenheit scale which people ordinarily know, obtaining an apparently bigger/lower figure. Scientists use Centigrade in measuring temperature. On this scale the lowest temperature actually achieved is $273.195°$ below zero—which is close to the lowest mark theoretically possible, approximately $-273.2°$. Molecular theory explains temperature in terms of the motion of molecules. At that theoretical limit molecules would have no thermal motion.

4. See also David M. Rorvik's article, "Artificial Inovulation: A Startling New Way To Have A Baby," *McCall's* (May 1969). There this author envisions at first anonymous panels of donors—couples, not single men or women only—to afford women the "therapy" of a fertilized egg. Then, perhaps only ten or fifteen years hence, according to Dr. E. S. E. Hafez, experimental biologist at Washington State University, a wife "can stroll through a special kind of market and select from one-day-old frozen *embryos,* guaranteed free of all genetic defects and described as to sex, eye color, probable IQ, and so on, in detail on the labels. Following the purchase, the embryo can be thawed and then implanted, under a doctor's supervision." Finally will come (according to Rostand) an era in which "children can be born whose parents have been long dead or separated by continents their entire lives." These are not only "mind-stretching" prospects. They will mean the destruction of parenthood as a basic form of humanity and thus the immeasurable dehumanization of man.

5. Dr. Heynes was endeavoring to counteract a lack of oxygen supply to the baby caused by the birth process itself. His procedure—which has *not* gained acceptance—was a treatment designed to be for the welfare of the small patient. It was not for the purpose of improving mankind.

6. The first woman to perform this reproduction on herself and to announce it to the world became a celebrity. Virgin birth cults sprang up

all over the country, and soon she was elected president of the United States. Here are some excerpts from her first Inaugural Address: "There is only one population problem. It is not the blind, the deaf, the feebleminded or the underprivileged. There is only one problem . . . those hairy, lewdniks with their puny, impotent organs and their second rate fertility. They are fit only for the lowest labor. Verily they will soon be redundant! Puny premature ejaculators. Never again will a Woman be forced to submit to their inept ministrations or to their heavy-handed rule . . . I give you the Danish Machine, a completely cybernated sexual system—sensual on command; vigorous on demand" (from the imagination of Professor John Batt, "They Shoot Horses, Don't They? An Essay on the Scotoma of One-Eyed Kings," *U.C.L.A. Law Review* 15 (Feb. 1968) 514-15).

7. "The New Biology," pp. 285, 281.

8. Ibid., p. 302.

9. Simultaneously Rosenfeld's book with the same title, *The Second Genesis,* was published by Prentice-Hall. The *Life* article was largely from one part of this remarkable book, on "The Exploration of Prenativity." The other two parts are "The Refabrication of the Individual" and "Control of Brain and Behavior."

10. Albert Rosenfeld, *The Second Genesis: The Coming Control of Life* (Englewood Cliffs, N.J.: Prentice-Hall), pp. 185-86; *Life* (June 13, 1969): 50.

11. Of course, abortion can never be a treatment for the child. To suppose so is to affirm that it can be therapeutic for a "life" to cause that life not to be.

12. Leroy Augenstein, *Come, Let Us Play God* (New York: Harper and Row, 1969), p. 116.

13. "Sex Control, Science and Society," *Science* 161 (Sept. 13, 1968): 1107-12; reported in *The New York Times,* Sept. 15, 1968.

14. By trying these techniques upon the unconceived child, I mean, of course, upon the sperm or ovum, or upon the genetic potentialities of the parents. I am aware that genetic surgery may be performed upon the zygote or the embryo after conception. That would clearly be a form of treatment and raises a different set of problems.

It must be noted, of course, that those who would proceed with genetic surgery at grave risk to the child they hope to produce are probably relying on intrauterine screening and abortion to remove the results of unsuccessful trials.

15. *Come, Let Us Play God,* pp. 103-04.

16. For an elaboration of these principles governing parental consent in behalf of a child postnatally, see Paul Ramsey, *The Patient As Person* (New Haven, Conn.: Yale University Press, 1970), chap. 1.

17. David L. Bazelon, "Medical Progress and Legal Process," *The Pharos* (Apr. 1969): 34.

18. Alfred Rosenfeld reports: " 'If I can carry a baby all the way through to birth *in vitro,'* says an American scientist who wants his anonymity protected, 'I certainly plan to do it—altho, obviously, I'm not going to succeed on the first attempt, or even the twentieth' " (*The Second Genesis,* p. 117).

19. For an example, see G. E. W. Wolstenholme, ed., *Man and His Future* (Boston: Little, Brown and Co., 1963), p. 290.

20. Donald M. MacKay in the discussion in *Man and His Future,* p. 286.

21. The subtitle of an article by Leonard J. Duhl, "Planning and Predicting," *Daedalus,* Toward the Year 2000 (Summer 1967): 779.

22. Frederick Winsor, *The Space Child's Mother Goose* (New York: Simon and Schuster, 1958, 1963), no. 30.

23. "Reflections on the New Biology," *U.C.L.A. Law Review* 15 (Feb. 1968): 269.

24. "The New Biology," p. 279. Obviously the risk of defective children is high for any half-brother and half-sister who unknowingly marry one another. At the present rate at which AID is practiced, however, the risk is low for the population as a whole. For this reason, some who do not disapprove of AID in itself would still disapprove of its widespread social acceptance which might follow from the legalization of the practice, because of the pedagogy of the law's enactment. For these and other societal and eugenic reasons, some believe concerning AID that "we should maintain strict principles (against its use), but be flexible in practice" in individual cases (Martin P. Golding, "Ethical Issues in Biological Engineering," *U.C.L.A. Law Review* 15 [Feb. 1968]: 472). The question of the morality of the practice, however, cannot be settled by statistically spreading the risk, or keeping it low. The risk is real; it is not negligible for the unborn; it is an *additional* risk. How, then, can the decision to do this be thought to be a responsible one?

25. "Ethical Issues in Biological Engineering," pp. 443-79, esp. pp. 451 ff.

26. Ibid., in order, pp. 453, 456 (cf 457), 464 (cf. 469).

27. Ibid., p. 470.

28. Joseph Fletcher, "Our Shameful Waste of Human Tissue," in Donald R. Cutler, ed., *The Religious Situation 1969* (Boston, Mass.: Beacon Press, 1969), p. 248.

29. Leroy Augenstein asks this sort of question "humbly, prayerfully, and above all responsibly," but I regret to say that the *title* of his book, *Come, Let Us Play God,* and defenses of this choice of title, equate any vital decision that is risk-filled for ourselves and others with playing a divine role. If every time a surgeon picks up a scalpel (or every time a minister influences people or a parent his children) he is "playing God"–or if a man and women in procreation are "playing God"–then, of course, the concept loses all meaning. It becomes merely an invocation to responsible action. This loose use of language must be given up if we are to search out the meaning of being men who know not to play God as a regulative norm in the "awesome business" that is before us (pp. 12, 146).

30. Karl Rahner, "Experiment: Man," *Theology Digest* 16 (Feb. 1968): 57-69; "Experiment Mensch: Theologisches über die Selbstmanipulation des Menschen," in *Schriften für Theologie* 8 (Einsiedeln, 1967): 260-85.

31. "Experiment: Man," p. 62.

32. Ibid., p. 61.

33. Ibid., in order, pp. 59, 60, 64. We shall see in a moment that there is a vague and largely unspecified "moral order" behind these statements of what "a truly alert morality" can demonstrate.

34. Ibid., in order, pp. 64, 57.

35. Ibid., p. 58.

36. Ibid., in order, pp. 59, 65, 60, 58. Indeed, scientifically directed self-creation and theology deal with the same reality: "man as a whole—man in the ultimate totality of his own self-awareness" (p. 57). Does this not make it evident that the God men will meet in the absolute future is the same god they now venture to be? Men now are the creators of the creativity they finally will become.

37. Ibid., p. 65.

38. Ibid., in order, pp. 66, 57, 64, 68.

39. Ibid., in order, pp. 61, 67, 68.

40. *The Ethics of Sex*, p. 199.

41. Donald Fleming, "On Living in a Biological Revolution," *The Atlantic Monthly* 223 (Feb. 1969): 64-70. Subsequent quotations, unless otherwise noted, are from this article.

42. In Wolstenholme, *Man and His Future*, p. 380.

43. A correction of Fleming's article may perhaps be in order here. The present writer has never said that "genetic tailoring" would necessarily be one of these self-violations of man by man. In fact, I specifically said the contrary (see above, pp. 44-45). I prefer the term "genetic *surgery,*" in order to keep clear the fact that the "patient" of medical care is the unborn child, and not that nonpatient, the human race, which can only be the object of artistic or mechanistic design and not of *care.* Above in this chapter (pp. 116-20) I have pressed the ethical analysis further, to ask what morally relevant circumstances can make "genetic surgery" the treatment of choice and what could *deny us this option.*

44. Gordon Rattray Taylor, *The Biological Time Bomb* (New York: World Publishing Co., 1968), p. 21.

45. Ibid., p. 231.

46. (Garden City, New York: Doubleday and Co., 1968). The idea of setting truly long-range goals for the human race as a whole has already been criticized above because unknown variables make questionable our human wisdom in setting these goals, and also because present and immediately future moral claims and right relations among men must take precedence over the unclear notion of an obligation to act now so as to bring about a remotely future state of affairs. Feinberg's definition of a long-range goal is "some desired future state of affairs whose realization would require an effort lasting over many generations" (p. 15). Men can *imagine* this to be the nature of human history, but no such direction of mankind is likely to be imposed, because it would require a guarantee that future generations will invariably adopt our present long-range projects for themselves, or build better ones *in continuity* with our present plans. Men are not likely to succeed in organizing their own Providence, although we are prone to dream that we can rationalize the flux of future events into something other than history.

47. Ibid., pp. 20, 21, 22, 25.

48. Ibid., p. 18.

49. Ibid., p. 31. "We bear the proud and lonely distinction of being the only known part of the world that has self awareness" (p. 50). Cf. the "thinking reed" passage in which Pascal celebrates the fact that man knows

that he dies while a vapor that kills him knows it not, and concludes that "man's whole dignity consists in thought" (*Pensées,* 347).

50. Ibid., p. 34.

51. Ibid., p. 43 (italics added).

52. Ibid., p. 47.

53. Ibid., p. 52.

54. Ibid., pp. 50-51 (italics added).

55. Ibid., p. 70.

56. Ibid., pp. 61, 74, 77.

57. Ibid., pp. 83, 85-87.

58. Ibid., pp. 63, 95, 93.

59. Ibid., pp. 104, 93.

60. See the account of "intracranial self-stimulation" in Rosenfeld, *The Second Genesis,* pp. 203-07.

61. *The Prometheus Project,* pp. 140, 141.

62. Ibid., pp. 142-43, 145.

63. Ibid., pp. 143, 145-46, 148.